ENGINEERING
OUR WAY
TO THE MOON

ENGINEERING OUR WAY TO THE MOON

Untold Apollo Stories

Richard Boudreau

ABOUT THE COVER

NASA Engineering Managers - at the Apollo 11 Liftoff

L-R, Charles Mathews, Wernher von Braun, George E. Mueller,
Samuel C. Phillips,   July 18, 1969. (NASA photo)

Charles Mathews, Deputy Associate Administrator, Office of Manned Space Flight (OMSF), he ran the Gemini Program.

Wernher von Braun, Director, Marshall Space Flight Center in Huntsville, Alabama. Responsible for developing the German V-2 rockets used during WW II, von Braun and his engineering team were brought to the United States and given responsiblity for development of all Apollo program launch vehicles.

George Mueller, OMSF Administrator, was NASA's top official for Manned Space Flight. He personally persuaded von Braun to use an "all-up" testing approach, in order to meet President Kennedy's moon landing schedule and to reduce runaway program costs.

Samuel Phillips, Apollo Program Director, working for Mueller he was key in implementing *Systems Engineering/Management procedures* within MSC, MSFC and KSC thereby successfully gaining control of Apollo program schedules and costs.

The significant contributions of these men to the success of the Apollo Program are described in "Who Accomplished this Feat" (pages 45–52), as well as their risky Apollo 8 "go" decision as described in "My Journal" (pages 105-108).

Cover Design by Aly Boccardi
aly.boccardi@gmail.com

Printed in the United States of America

First Edition 2019
10 9 8 7 6 5 4 3 2 1

TO WILMA

Your enthusiasm, energy and unbounded caring
make all the good things happen in my life

CONTENTS

CONTENTS - continued

ABBREVIATIONS and ACRONYMS

ABMA	Army Ballistic Missile Agency
AF	Air Force
AFSC	Air Force Systems Command
AGC	Apollo Guidance Computer
AGS	Abort Guidance System
AOT	Alignment Optical Telescope
APS	Ascent Propulsion System
ASPO	Apollo Spacecraft Program Office
ASSCCB	Apollo Spacecraft Software Configuration Control Board
CCB	Configuration Control Board
CDH	Constant Delta Height
CIRC	Circularization
CM	Command Module
CSI	Coelliptic Sequence Initiation
CSM	Command Service Module
CSOC	Consolidated Space Operations Center
DOD	Department of Defense
DOI	Descent Orbit Insertion
DPS	Descent Propulsion System
DSKY	Display Keyboard Assembly
EI	Entry Interface
EOR	Earth Orbit Rendezvous
EVA	Extra-Vehicular Activity
FDAI	Flight Director Attitude Indicator
FOD	Flight Operations Directorate
GAEC	Grumman Aerospace Engineering Corporation
GNC	Guidance Navigation and Control
GSFC	Goddard Space Flight Center
IBM	International Business Machines

ICBM	Inter-Continental Ballistic Missile
ICD	Interface Control Document
IMU	Inertial Measurement Unit
JPL	Jet Propulsion Laboratories
JSC	Johnson Space Center
KSC	Kennedy Space Center
LEO	Low Earth Orbit
LM	Lunar Module
LOI	Lunar Orbit Insertion
LOR	Lunar Orbit Rendezvous
MCC	Mid Course Correction
MCC-H	Mission Control Center-Houston
MIT	Massachusetts Institute of Technology
MOL	Manned Orbital Laboratory
MPAD	Mission Planning and Analysis Division
MSC	Manned Spacecraft Center
MSFC	Marshall Spaceflight Center
MSFN	Manned Space Flight Network
NAA	North American Aviation
NASA	National Aeronautics and Space Administration
NR	North American Rockwell Corp.– Space Division
OMSF	Office of Manned Space Flight
PDF	Portable Document Format
PERT	Program Evaluation Review Technique
PGNCS	Primary Guidance, Navigation and Control System
RAM	Random Access Memory
RCS	Reaction Control System
REFSMMAT	Reference to Stable Member Matrix
ROM	Read-Only Memory
RR	Rendezvous Radar
RTCC	Real-Time Computer Complex
R-W	Ramo-Wooldridge

SETD	Systems Engineering and Technical Direction
SM	Service Module
SPS	Service-Module Propulsion System
STL	Space Technology Laboratories
TDRSS	Tracking and Data Relay Satellite System
TEI	Trans-Earth Injection
TLI	Trans-Lunar Injection
TPI	Terminal Phase Initiation
TRW	Thompson-Ramo-Wooldridge
UHCL	University of Houston Clear Lake
WBS	Work Breakdown Structure
ΔV	Spoken as "delta V" – an incremental spacecraft velocity change, measured in feet per second accomplished by a rocket engine "burn" that places the spacecraft onto a new trajectory/orbit.

PREFACE

My purpose in writing this book is to fill a space history void – TRW's significant role in the development of Apollo Mission Techniques. This is a story that is missing in today's literature, either published or online. Part III of this book is that story

This book is in three parts:

Part I – The Space Race

Part I is included to provide historical context relevant to the Apollo Mission Techniques story. This includes an overview of the space race, how it started and its early beginnings when the Soviets were clearly in the lead.

Next an overview of America's race to the moon; the Apollo launch vehicle and spacecraft configurations, trajectories flown, TRW's role, and NASA organizations and key individuals that made it happen – especially George Mueller, Samuel Phillips and Wernher von Braun.

Finally the Soviet space programs during those early times and their significant moon landing efforts. After glasnost, we now know it was a real race to the moon.

Part II – My Journal – a trip to the Moon

Part II covers how I came to be working on Apollo Mission Techniques in Houston - a personal recounting of the first thirteen years of my engineering career after college, working on GNC systems in Huntsville, San Diego and Houston supporting America's man-in-space efforts.

Included is how I first learned of TRW's pioneering systems engineering work for the Air Force that led to my joining TRW in Houston to work on America's moon landing program.

Personal events such as meeting and marrying Wilma and the birth of our two children are also here. It was much fun and good times for my family during our 9-year Houston adventure.

An overview of the Apollo Mission Techniques activity and my role therein is also here – a definite high point in my engineering career.

Part III – Apollo Mission Techniques – A TRW story

Part III was written for space historians and enthusiasts - a detailed description of the early beginnings of the Apollo Mission Techniques effort, the what, why and how this important activity was initiated by TRW during the Gemini program, and then after the Apollo fire how Howard W. "Bill" Tindall Jr. with continuing TRW support carried out this effort until the first moon landing. Mr. Tindall is widely recognized for his significant contributions to the success of the Apollo program, with his many large meetings and "Tindallgrams".

Part III describes how TRW, in conjunction with Mr. Tindall's working groups, produced all the mission techniques documents that defined the recognized baseline plan for trajectory control for all manned Apollo missions.

ACKNOWLEDGMENTS

Thanks in writing this book are due to:

Marvin Fox – the "inventor" of Apollo Mission Techniques. My direct supervisor, Marvin first recognized the problem, was key in developing the "data priority/decision logic" approach to its solution and effectively advocated for its adoption by NASA/MSC management.

Howard W. "Bill" Tindall, Jr. – as Chief, Apollo Data Priority Coordination, Bill was my NASA/MSC manager for TRW's Mission Techniques effort, a two-plus year activity to ensure the vast Apollo team worked to a recognized GNC baseline for how the missions were to be flown.

Dr. George Mueller – while writing this book, I learned that Dr. Mueller had left TRW to join NASA as head of the Office of Manned Space Flight (OMSF) only if the directors of MSC, MSFC and KSC reported directly to him.

This was done, and his effective implementation of systems engineering/system management principles at MSC very likely smoothed the way for the Apollo Mission Techniques effort and for the documents to become a controlled baseline, a position I advocated with Tindall who made it happen.

Bob Kidd – with me from the beginning, Bob worked Mission Techniques for the critical Lunar Descent Phase.

Dennis Greene, a much-loved uncle - for his insistence over the years that I write my moon landing story for our family's enjoyment.

My Children - Michele and Roland - for coming with me on my nostalgia trip. Hopefully in the distant future this book will help them decide what to do with all the Apollo stuff in my file cabinet.

My Grandchildren - Ben & Anna - for being a part of my wonderful family. I am proud of you both for your successful careers in Silicon Valley and hope you enjoy this book.

Ku'u– your moon landing skepticism gave me another reason to write this. When you see how many individuals were involved, you'll agree it would have been literally impossible to do and keep secret a faked moon landing.

and of course Wilma – my Hawaiian Portuguese Princess wife of 55+ years - for your encouragement, support and tolerance of the many hours I spent at my computer writing this account.

ABOUT THE AUTHOR

Mr. Boudreau's engineering career started during the early days of the space race; first working with Wernher von Braun's team in Huntsville when they placed the first US satellite into earth orbit, then with General Dynamics Astronautics working Atlas-Centaur Guidance and Control, then with TRW in Houston on the Apollo Moon Landing Program. Later in Redondo Beach, California, he managed an Electronic Systems Design Department and then held Project Management positions for TDRSS, CSOC and several classified projects. He is now retired and living in Hawaii.

He has a BS in Electrical Engineering from the University of Michigan and MBA in Business and Management from Pepperdine University.

While working at TRW Houston Operations he managed TRW's Apollo Mission Techniques effort, from its first beginnings in early 1966 until the first lunar landing in July, 1969 - first for Dr. Shea, ASPO manager, and then for Bill Tindall, ASPO Chief of Apollo Data Priority Coordination.

Boudreau's TRW team authored and published the approximately 50 Apollo Mission Techniques documents that are now housed in the University of Houston Clear Lake (UHCL) Archives, representing the combined efforts of many individuals working within the engineering teams organized by Mr. Tindall.

In October 2015 Mr. Boudreau donated to the UHCL Archives three documents from the earlier work done under Dr. Shea:

"TRW Rendezvous Study" - a Gemini post-flight analysis report that identified problems in the priority and use of alternate GN&C data sources.

"Apollo Mission Techniques" – An Overview document including the chapter for lunar ascent/rendezvous and placeholder sections for the other mission phases.

"Apollo Mission Techniques" - lunar descent techniques chapter.

A finding aid for these three documents is available at:

https://uhcl-ir.tdl.org/uhcl-ir/bitstream/handle/10657.1/411/20150009.pdf?sequence=1&isAllowed=y

INTRODUCTION

Have you ever looked up at the moon and wondered just how we flew there (and back) in 1969 – almost 50 years ago?

The huge Saturn V launch vehicle boosted the two Apollo spacecraft - one specifically for landing on the moon, the other with three Astronauts inside - first into a circular earth orbit and then onto a trajectory aimed roughly towards the moon, 240,000 miles and three days away. After that, however, how did we control the *trajectories* of these spacecraft such that:

- they neither missed nor impacted the moon, but went instead into a safe orbit around the moon?

- And once in Lunar orbit, how did we separate the LM from the CSM "mother ship" and place the LM into a much lower but still safe orbit around the moon - in preparation for actually landing on the moon . . . .

  – and then, how did we safely accomplish the final powered descent to the moon's surface – not too fast or steep so we didn't crash onto the moon, but not too slow and careful either, which would result in propellant depletion, thereby also leading to crashing onto the moon?

- And once safely on the lunar surface, how did we prepare the LM GNC systems for launch into a safe orbit around the moon?

  – and then how did we accomplish the orbital maneuvers needed to successfully rendezvous and dock with the CSM "mother ship"?

- Then, after the LM crew transferred back into the CSM and the LM was jettisoned, how did we inject the CSM onto a safe return-to-earth trajectory

  – such that the Command Module part of the CSM would re-enter the earth's atmosphere safely - not too steep where it would burn up, or not too shallow where it would skip-out into an unsafe orbit or land at an unknown spot somewhere on earth?

- And for all of these trajectory control maneuvers, how did we do the tracking, navigation, ΔV targeting computations and data exchanges needed between ground mission control and the onboard Guidance, Navigation and Control (GNC) systems and astronauts such that they could successfully accomplish the propulsive rocket burns (ΔV's), many times behind the moon and out of communications with the flight controllers on earth?

- And how did we know that the various onboard and ground GNC systems were all operating correctly and accurately so we could safely continue and not have to abort or re-plan the mission? - in particular, if the various GNC data disagreed beyond acceptable levels, then what? And what were those acceptable limits of disagreement for the various times in the mission?

The answers to these questions and more are contained in 12 key documents, produced over a 2-year period before the first moon landing. They are called Apollo Mission Techniques.

This book provides a behind-the-scenes look at the extensive activity that created these documents - how TRW first identified their need, then worked with NASA to launch the "data priority" activity that defined and documented a controlled baseline for how the many ground and onboard components of the GNC systems were to be used for the Apollo missions.

I was the lead TRW engineer for this effort, first during their initial development under Dr. Shea and then under Bill Tindall up through the time of the first lunar landing in July 1969. Working with the various mission phase working groups set-up by Tindall, my TRW team wrote and published all of the Apollo Mission Techniques documents in the University of Houston Clear Lake archives – about 50 different documents covering all mission phases for all the manned Apollo missions.

WHY I WROTE THIS BOOK

My interest in the Apollo moon-landing program and Space History in general was re-kindled in March 2014 during my 80th Birthday "Nostalgia Trip" to Huntsville, Alabama and Houston, Texas.

Our first stop was Huntsville, where I had worked with von Braun's engineers while in the Army's Ballistic Missile Agency (ABMA). We visited the U.S. Space and Rocket Center museum and toured Marshall Space Flight Center/Redstone Arsenal – we saw many rockets, much space hardware, and interesting exhibits reminiscent of my Army Huntsville experience. We saw too an actual Saturn V launch vehicle left-over from when the Apollo program was cancelled earlier.

At Houston's Rocket Park - another left-over Saturn V. Other Houston highlights included a tour of the old Apollo Mission Control Center, a luncheon with eight of my former TRW co-workers from over 40 years ago and visits with co-workers and friends still residing near Houston. All-in-all a wonderful trip - made especially enjoyable as these experiences were shared with my wife Wilma, our children Michele and Roland, Michele's husband John, and our granddaughter Anna.

After the trip I searched many internet sites for *Apollo Data Priority, Apollo Mission Techniques, TRW* Houston Operations, Tindall, etc. – relevant to my role on the Apollo program. There was much information on Bill Tindall and his unique "Tindallgrams" but only a few references on the overall scope of Mission Techniques/Data Priority, and no information at all on TRW's role in their development - hence this book.

I did find in the University of Houston Clear Lake (UHCL) archives many of the Apollo Mission Techniques documents – NASA documents that were authored and published by TRW under my direction. Thank you, Jean Grant at UHCL, for providing me PDF copies of those that directly governed the first lunar landing – a nice supplement to my file of original Apollo documents.

I was fortunate to have participated in mankind's first venture to another celestial body and very much enjoyed revisiting memories from those earlier days while writing this book.

But my journey is not yet over. The 50th anniversary of the first moon landing will be July 20, 2019, and by happy coincidence the eleventh Napoleon Boudreau Reunion in Upper Michigan will be held mid-July 2019. My plan is to stop by Houston during this trip to again visit old friends in the Houston area and possibly attend any moon-landing celebratory events.

MY NOSTALGIA TRIP

HUNTSVILLE - REDSTONE ARSENAL

Redstone Test Site

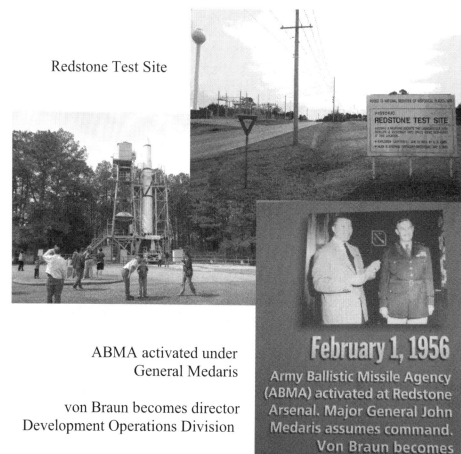

ABMA activated under
General Medaris

von Braun becomes director
Development Operations Division

February 1, 1956

Army Ballistic Missile Agency
(ABMA) activated at Redstone
Arsenal. Major General John
Medaris assumes command.
Von Braun becomes
                                pment
                                vision.

L to R Ernst Stuhlinger • Helmut Hoelzer • Karl Heimburg • Ernst Geissler • Erich W. Neubert • Walter Haeussermann
Wernher von Braun • William Mrazek • Hans Hueter • Eberhard Rees • Kurt Debus • Hans H. Maus

HUNTSVILLE – U.S. SPACE & ROCKET CENTER

**NASA Marshall
Space Flight Center**
Visitor's Entrance

**a leftover
Saturn V**

HOUSTON

Apollo Mission
Control Center
Circa 1969

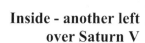

Inside - another left
over Saturn V

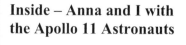

Inside – Anna and I with
the Apollo 11 Astronauts

A Mini-Reunion in Houston of former TRW colleagues
(from TRA's Gazette – May/June 2015)

To celebrate his 80th birthday in March, 2014, Dick Boudreau took his family on a nostalgia trip back to his old stomping grounds in Houston, TX and Huntsville, AL. Dick and his wife now live in Kailua, HI. His former colleagues at Houston arranged a great reunion visit with old friends from TRW Houston Operations.

Here they are at Frenchie's Italian Restaurant near the Johnson Space Center.

L-R - Darwin Miller, Dick Boudreau, Bob Rountree, Bob Kidd, & Marv Fox.

L-R - Bob Kidd, Marv Fox, Dick Boudreau, Gary Dinsmore, Miguel Zamora, Jim Duke, Darwin Miller, Bob Rountree, Roscoe Lee

HOUSTON – A SIDE TRIP TO CAJUN COUNTRY

Our traveling Group - Prejean's Cajun Restaurant in Lafayette, LA
L–R   Richard, Wilma, Roland, Michele, John, Anna

Cajun Country Swamp Tour
Breaux Bridge, Louisiana
A cold morning!

PART I

THE SPACE RACE

In 1955, both the United States and the Soviet Union were building ballistic missiles that could be utilized to launch objects into space with enough velocity to orbit the earth.

On July 29, 1955, President Eisenhower's press secretary announced that the United States intended to launch "small Earth circling satellites" sometime between July 1, 1957 and December 31, 1958 as part of their contribution to the International Geophysical Year.

One month later, on August 30, 1955, the Soviets responded; The Soviet Academy of Sciences created a commission whose purpose was to beat the Americans into Earth orbit - *the de facto start date of the Space Race.* Sergei Pavlovich Korolev was the individual most credited for this action.

Two years later, on October 4, 1957, the Soviets launched the first ever artificial satellite into earth orbit – Sputnik 1, weighing 184 pounds. "Sputnik" is the Russian word for "Traveler."

And one month after that, on November 3, 1957, Sputnik 2, weighing 1,118 lbs., was launched with the doomed canine Laika aboard - *the first living being in orbit*, an even more alarming event to the Americans.[1]

The space race was on in earnest - ten years after the start of the Cold War. Sputnik 1 circled the earth once every 90 minutes, and the Soviets had also put

---

[1] Liaka the dog had provisions for one week; Sputnik 2 burned up upon reentry after 162 days in orbit.

a living being into earth orbit, spawning fears that they had leapt to technological and military superiority in a new and dangerous nuclear age. Both were continuing development of their nuclear tipped missiles; but now there was the possibility of space-based atomic weapons.

In response, President Eisenhower ordered project Vanguard to move up its timetable and launch its satellite much sooner than originally planned. However, the December 6, 1957 launch at Cape Canaveral, broadcast live on US television, was a monumental failure. Exploding a few seconds after launch, the Vanguard program became an international joke, the satellite described in newspapers as Flopnik, Stayputnik, Kaputnik and Dudnik. At the United Nations, the Russian delegate offered the US. Representative aid "under the Soviet program of technical assistance to backwards nations."

Only in the wake of this very public failure did von Braun's Redstone team get the go-ahead to launch the Explorer 1 satellite using their Jupiter-C rocket developed by the Army Ballistic Missile Agency (ABMA). Four months later, on Jan. 31, 1958, they successfully launched America's first artificial satellite, Explorer 1, with a payload weight of 31 pounds, a somewhat anemic accomplishment when compared to the Sputnik payloads of 184 and 1,118 pounds.

The Soviets launch the first man into earth orbit

Three years later, on April 12, 1961, the Soviets launched the first *man* into space – Yuri Gagarin's one-orbit flight in Vostok 1. This was a seminal event, noticed in particular by America's Space Task Group who, one month later, on May 5, 1961, launched the first flight of our Mercury man-in-space project, Alan Shepard's *15 minute sub-orbital* flight.

The Soviets had flown at least three successful flights that monitored dogs in orbit before they considered it safe to send a man into space and survive in zero gravity. Sputnik 10 was a full test version of the Vostok human-carrying space capsule, and its canine passenger was successfully recovered before Yuri Gagarin's orbital flight.

But it wasn't until 10 months later, in February 1962, that by using a more powerful Atlas booster rocket we also launched a man into earth orbit – John Glenn's 3-orbit flight.

Americas moon-landing declaration

Only 20 days after Alan Shepard's sub-orbital flight, President Kennedy on May 25, 1961, announced our nation's goal of landing a man on the moon before the end of the decade.

A gutsy move in the face of major unknowns, questions with no answers at that time:

- Can humans safely exist in Zero-gravity for the 14 days needed for a moon trip?

- How do we provide for astronauts life support outside an enclosed space capsule?

- How should we go to the moon? It would require an extremely large spacecraft and launch vehicle, or to reduce the size of either the launch vehicle or the spacecraft, one could do rendezvous/docking of smaller vehicles, either in earth orbit or in lunar orbit.

- How do we rendezvous two spacecraft in earth orbit, or worse, around the moon?

- Can the moon's crust support a soft landing? How deep is the dust on the moon?

- What is the moon's largely unknown topography and gravitational field?

President Kennedy's announcement in spite of these uncertainties vastly changed the playing field, resulting in the Soviets redirecting their manned and unmanned space programs as needed towards also landing a man on the moon.

The results of both the Soviet and American space programs are summarized in the following table – a timeline of the major American and Soviet programmatic events on the way to the moon.

Space Race Milestones

		SOVIET	USA

Oct 4, 1957	**First artificial earth satellite launched - "Sputnik 1" satellite weight 184 lbs.**
Nov 3, 1957	**First living being into orbit "Laika" canine- "Sputnik 2" satellite weight 1,118 lbs.**
Jan 31, 1958	First US satellite - launched by U.S. Army Redstone-Jupiter-C. Explorer 1 weight 31 lbs.
Sep 14, 1959	**First spacecraft to impact the moon - 854 lbs.**
Jan 20, 1961	President Eisenhower's two terms end. Kennedy becomes President.
Apr 12, 1961	**First man in space - Yuri Gagarin's one-orbit flight in Vostock-1.**
May 5, 1961	First American in space – Alan Shepard's Mercury-Redstone sub-orbital flight.
May 25, 1961	**President Kennedy - states US goal to land man on the moon by end of the decade.**
Feb 20, 1962	First of 4 Mercury-Atlas orbital flights - John Glenn's 3-orbit flight.
Jul 11, 1962	NASA - James E. Webb announces Apollo mission mode is Lunar Orbit Rendezvous.
May 15, 1963	Last of 4 Mercury-Atlas flights. 22 orbits.
Jun 16, 1963	**First woman in space – Valentina Tereshkova's three-day mission in Vostock-6.**
Sep 3, 1963	**Mueller heads OMSF. NASA reorg: MSC, MSFC & KFC report directly to Mueller.**
Nov 22, 1963	President Kennedy assassinated. Lyndon Johnson becomes President.
Oct 14, 1964	**Leonid Brezhnev becomes leader of Soviet Union - until 1982. Khrushchev retires.**
Mar 18, 1965	**First spacewalk - by Alexi Leonov in Voskhod 2.**
Mar 23, 1965	First of 10 Gemini flights. First American spacewalk June 3, 1965 - Gemini 4.
Feb 3, 1966	**Luna 9 - First spacecraft to soft-land on the moon**
Mar 31, 1966	**Luna 10 - First spacecraft to orbit the moon.**

May 30, 1966	Surveyor-1 - First US spacecraft to soft-land on the moon.
Aug14, 1966	Lunar Orbiter-1 - First US spacecraft to orbit the moon.
Nov 11-15, 1966	Last Gemini flight.
Jan 27, 1967	**Apollo 1 fire - Grissom, White, Chaffee died on the launch pad in the Block 1 CM.**
Apr 23-24, 1967	**Soyuz 1 Tragedy - Cosmonaut Vladimir Komarov died - main parachute failed on reentry - descent module impacted earth at 89 mph and burned.**
Nov 9, 1967	**Apollo 4/AS-501 - First test flight of Saturn V, "all-up" unmanned. SPS used to achieve lunar return velocity for successful heat shield test. Saturn V - nearly perfect.**
Jan 22, 1968	Apollo 5/AS-204 - First unmanned test of LM in space with Saturn IB. A successful 7-orbit mission.
Apr 4, 1968	**Apollo 6/AS-502 Second Saturn V test flight** – an unmanned, "all-up" mission. Pogo damage prevented S-IVB re-start, poor heat shield test.
Sep 14-21, 1968	**Zond 5 - First Spacecraft to loop around the moon and return to earth with live test subjects.**
Oct 11-22, 1968	Apollo 7/AS-205 - First manned Apollo flight. First test of Block II CSM in LEO. Saturn IB booster.
Oct 25-30, 1968	**Soyuz 2 & 3 - dual LEO mission.** Soyuz 3 manned launch for rendezvous & docking with unmanned Soyuz 2. Rendezvous successful, docking failed.
Dec 21-27, 1968	**Apollo 8/AS-503 First humans to orbit the moon and return safely to earth. A spectacular success. Saturn V's third launch almost perfect, pogo OK.**
Jan 14-18, 1969	**Soyuz 4 & 5 dual LEO mission.** Soyuz 4 crew of 1, return with 3; Soyuz 5 crew of 3, return with 1.
Feb 21, 1969	N-1 test mission #1 – failed due to catastrophic launch vehicle failure shortly after lift-off.
Mar 3-13, 1969	Apollo 9 - First manned flight of LM - in earth orbit. LM rendezvous/dock with CSM.

May 18-26, 1969	Apollo 10 - Lunar Orbit & Return. "Dress Rehearsal" mission to confirm systems readiness.
Jul 4, 1969	N-1 test mission #2 - catastrophic launch vehicle failure shortly after lift-off.
Jul 16-24, 1969	**Apollo 11 - First humans to walk on the moon and safely return to earth. Neil Armstrong and Edwin "Buzz" Aldrin walked on the moon.**
Oct 11-18, 1969	Soyuz 6, 7 & 8 – triple LEO mission with crew EVA exchanges after docking. Successful.
Nov 14-24, 1969	Apollo 12 - Lunar Landing & Return. Pete Conrad and Alan Bean walked on the moon
Apr 11-17, 1970	Apollo 13 - Aborted mission - lunar flyby & return. Astronauts Lovell, Swigert & Haise.
Jun 1-19, 1970	**Soyuz 9 – LEO, new space duration record of 18 days. Gemini 7 was 14 days.**
Jan 31-Feb 9, 1971	Apollo 14 - Lunar Landing & Return. Alan Shepard and Ed Mitchell walked on moon.
Apr 23-25, 1971	Soyuz 10 – failed to dock with the Salyut 1 Space Station launched April 19, 1971.
Jun 7-Jun 30, 1971	**Soyuz 11 Tragedy.** The crew occupied the Salyut 1 Space Station 23 days. Before reentry, the Crew Capsule depressurized, killing the 3 cosmonauts. **The only humans known to have died in space.**
Jun 27, 1971	N-1 test mission #3 – catastrophic launch vehicle failure shortly after lift-off.
Jul 26-Aug 7, 1971	Apollo 15 - Lunar Landing & Return. David Scott and James Irwin walked on moon.
Apr 16-27 1972	Apollo 16 - Lunar Landing & Return. John Young and Charles Duke walked on moon.
Nov 23, 1972	**N-1 test mission #4** – catastrophic failure shortly after lift-off – a final launch attempt for their moon program. Soviet moon program suspended. Soviets re-focus to earth-orbiting Space Stations.
Dec 7-19, 1972	Apollo 17 – Lunar Landing & Return. Cernan and Schmitt walked on moon. **The Last Apollo Mission**
Sep 27-29, 1973	Soyuz 12 – first successful flight of the re-designed Soyuz spacecraft following the Soyuz 11 tragedy. First of over 100 more successful Soyuz flights.

THE AMERICANS IN THE SPACE RACE

How should we go to the moon?

In the early planning stages for Apollo, three different approaches to landing a man on the Moon were considered: direct ascent, rendezvous in Earth orbit, and rendezvous in lunar orbit.

For direct ascent, a huge monolithic rocket called "Nova" would be used to send a spacecraft on a straight path from the Earth to the Moon. The spacecraft would land and re-launch directly from the Moon - making it very heavy, primarily because of the increased weight of the propulsion systems needed to land on and launch from the moon - a much heavier command/service module with heat shield and life support systems for the two-week mission.

The second mode was earth orbit rendezvous (EOR), where two smaller rockets would assemble the same heavy spacecraft in orbit around the Earth. The astronauts would rendezvous with the fuel tank in Earth orbit and depart for the Moon. This method was initially favored by von Braun's team – envisioning this as a forerunner to space station operations in earth orbit.

The third mode - Lunar Orbit Rendezvous (LOR) was the method eventually decided upon. For this mode, a single rocket launches both the Command/Service Module (CSM) and the Lunar Module (LM) on a trajectory toward the moon. When the combined spacecraft reaches lunar orbit, one of the three astronauts remains with the CSM; the other two go into the LM, undock from the CSM, and descend onto the surface of the Moon. They then use the LM ascent stage to rejoin the CSM in lunar orbit, and use the CSM only for the return to Earth.

LOR significantly lightened the spacecraft and therefore simplified the launch vehicle. Although more complex from a trajectory control viewpoint – clearly rendezvous and docking (in lunar orbit yet) were of critical importance.

The LOR approach also brought added safety to the mission by providing variable length stopping points in earth orbit and lunar orbit as needed to resolve problems. Then ΔV maneuvers (rocket motor "burns" to change the trajectory) could be done to either continue the mission or bring the crew home from Earth or Lunar orbit - much easier than from a direct lunar transit.

At the Lunar Mode Decision Conference in June 1962, at the end of the day-long meeting, Wernher von Braun stood up and read from a sheet of paper he had been scribbling on:

> "We at the Marshall Space Flight Center readily admit that when first exposed to the proposal of the Lunar Orbit Rendezvous Mode we were a bit skeptical—particularly of the aspect of having the astronauts execute a complicated rendezvous maneuver at a distance of 240,000 miles from the earth where any rescue possibility appeared remote. In the meantime, however, we have spent a great deal of time and effort studying the [three] modes and we have come to the conclusion that this particular disadvantage is far outweighed by [its] advantages."

NASA announced the decision to pursue the lunar-orbit-rendezvous strategy at a press conference on July 11, 1962 – 13 months after Kennedy had announced his moon landing goal.

The Mercury Program

Project Mercury was the United States' first man-in-space program.

The one-man Mercury capsule was first launched on May 5, 1961 - a flight that was sub-orbital due to the propulsive limits of the Mercury-Redstone booster. This flight lasted 15 minutes and put the first American into space - Alan Shepard.

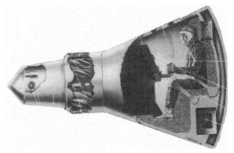

A second sub-orbital flight was done two months later.

The next four Mercury flights used a more powerful rocket - the Mercury-Atlas booster, which could launch the spacecraft into earth orbit. The last Mercury flight was 22 orbits on May 15-16, 1963.

The six successful flights of the Mercury Program proved man's ability to survive in space, operate a spacecraft, and perform experiments.

When I was in Huntsville, I was project engineer for flight testing the Mercury capsule's horizon scanners – as described in Part II of this book.

The Gemini Program

Gemini was the second NASA human spaceflight program.

A robust program, there was a total of ten manned flights in 1965 and 1966.

The Gemini two-man spacecraft was launched into earth orbit by a modified Titan II ICBM.

For the rendezvous and docking missions, a separate Atlas-Agena was launched as an on-orbit target, the Agena stage having been fitted with a docking target.

The Gemini program successfully accomplished its three main goals:

1) do space missions long enough for a trip to the Moon and back – 14 days,
2) conduct extravehicular activities (EVA) - working in space outside a spacecraft, and
3) develop in Low Earth Orbit (LEO) the orbital maneuver techniques as needed for rendezvous and docking operations around the moon.

With these new techniques proven by Gemini, Apollo could pursue its prime mission without doing these fundamental exploratory operations.

The Gemini missions used the newly built Mission Control Center at Houston for flight control, and all missions were successfully flown. The Gemini spacecraft was robust enough that the United States Air Force initially planned to use it for their Manned Orbital Laboratory (MOL) program (later canceled).

TRW's post-flight analysis of Gemini flight results as applicable to the Apollo program gave rise to the Apollo Mission Techniques activity - as described in Part II (briefly) and in Part III of this book.

The Apollo Program

It happened July 20, 1969 – after over 2,000 years of recorded history, man for the first time left the planet Earth to successfully fly to and tread upon the surface of another celestial body – a singular milestone in the history of the human race.

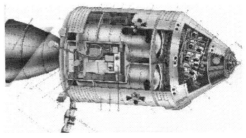

APOLLO CSM "MOTHER SHIP"

The short eight-year span it took to accomplish this feat is considered by many to also be one of the greatest engineering accomplishments of the 20th century. Six lunar landings (Apollo 11, 12, 14, 15, 16, and 17) returned a wealth of scientific data and 842 pounds of lunar samples.

APOLLO LM – ON THE MOON

Many thousands of persons were responsible for the success of this program, requiring the coordinated efforts of many NASA Centers and contractors to design and build the launch vehicles and spacecraft, the ground network for worldwide tracking of the spacecraft, and then to develop all the plans, procedures, software and training needed for ground and flight crews to operate and maintain all this equipment, and then to successfully fly the two space vehicles to the moon and back. Needed also were the management systems to control these activities, including feedback between the development organizations to identify needed changes to the systems such that performance, cost and schedule requirements were met - an immense management as well as systems engineering challenge.

And – all this was accomplished using electronic and computer technologies that were ancient in comparison to today's capabilities. The onboard computers, the CM and LM Primary Guidance Computers (AGC & LGC) and the computer for the LM Abort Guidance System (AGS) all used small magnetic cores for data storage. Each ferrite core stored only one bit of information (a 0 or 1).

The AGC with its 16-bit word-length had 2048 words of RAM and 36,864 words ROM; about 80,000 bytes total. Compare this to today's typical 16 GB memory stick with 16,000,000,000 bytes storage; the memory stick is essentially weightless compared to the AGC's 70 pound weight.

The ground-based Mission Control Center (MCC) computers used punched cards and magnetic tape for data input and storage - typically IBM System 360's and 7094's.

For engineering analyses and simulation, we were also in the stone age of computing where typically one run per day was the standard - submit your punched cards one day and get a printout of your results the next morning. At that time a few tabletop teletype machines were available which one could use to call up (using a modem) a time-share computer in downtown Houston over the telephone line and run fairly simple FORTRAN or BASIC programs and get your results right away - printed by the teletype machine – very slow in comparison to today, but great to have that much computing power with results immediately available.

APOLLO OVERVIEW

Major parts of the Apollo systems and how they were used in flying to the moon and back are shown here; also basic information useful in understanding how this engineering feat was accomplished.

THE LAUNCH CONFIGURATION

THE SPACECRAFT – CSM & LM

A NOTE ON SPACECRAFT ORBITS

MISSION TRAJECTORIES

LUNAR RENDEZVOUS TRAJECTORIES

GNC TERMS

THE LAUNCH CONFIGURATION

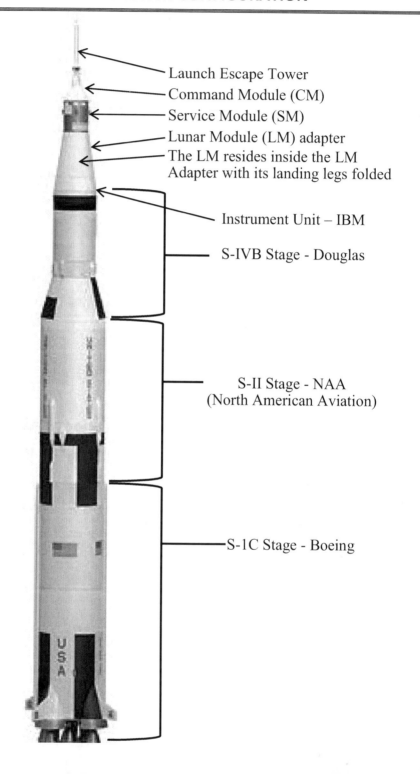

Launch Escape Tower

Command Module (CM)

Service Module (SM)

Lunar Module (LM) adapter

The LM resides inside the LM
Adapter with its landing legs folded

Instrument Unit – IBM

S-IVB Stage - Douglas

S-II Stage - NAA
(North American Aviation)

S-1C Stage - Boeing

THE LAUNCH CONFIGURATION

The Saturn V launch vehicle is a four stage booster designed to place the Apollo spacecraft onto a translunar trajectory. There are three powered stages. A fourth unpowered stage, the Instrument Unit (IU), contains Saturn V's GNC system.

A normal launch consists of a boost into Earth Parking Orbit (EPO) by burning practically all the S-IC and S-II stage propellants and partially burning the S-IVB propellant. A second S-IVB burn, the translunar injection (TLI) burn, is done after 2-4 revolutions in the EPO.

Shortly after the TLI burn, the CSM "mother ship" separates from the LM adapter and performs "transposition and docking" - the CSM separates from the S-IVB, rotates and docks with the LM, followed by extraction of the LM from the S-IVB. The crew then enters the LM for checkout of its systems.

Shown below is the spacecraft configuration during its three-day coasting flight to the moon.

Only the CM with its crew of three astronauts returns to earth after completion of the lunar landing portion of the mission.

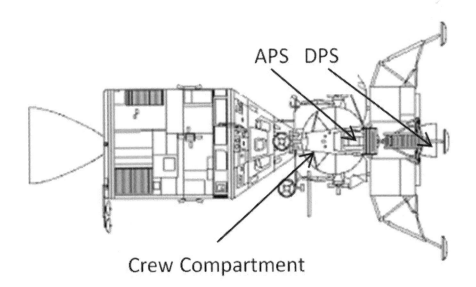

APS DPS

Crew Compartment

CSM shown docked with the LM

THE COMMAND AND SERVICE MODULES

CM RCS Thrusters

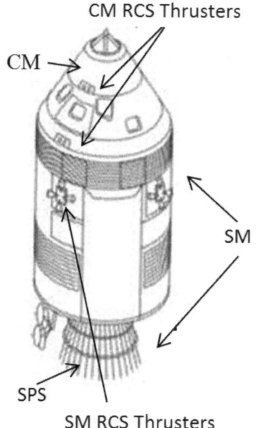

CM

SM

SPS

SM RCS Thrusters

CM---Command Module. The cone-shaped capsule within which the three astronauts live during most of their 7 to 10 day mission. The CM contains the Primary Guidance, Navigation and Control System (PGNCS) and a Reaction Control System (RCS) used for attitude control of the CM.

SM--Service Module

The SM contains the Service Module Propulsion System (SPS), the main rocket engine used for the large LOI and TEI ΔV maneuvers.

The SM also has a separate RCS for smaller ΔV maneuvers and attitude control of the CSM.

The SM also contains electric power & life support systems.

CSM--The combined CM and SM The CSM is also referred to in this book as the "mother ship" when separated from the LM during the lunar orbit, landing and rendezvous mission phases.

THE LUNAR MODULE

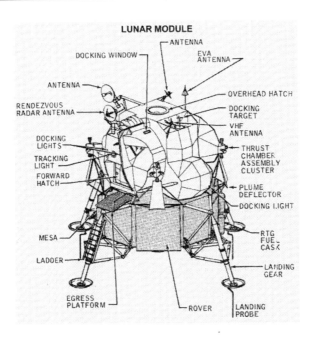

LM   Lunar Module - pronounced "lem"- the two-man spacecraft that is launched with the CSM into the trans-lunar trajectory.

A two-stage vehicle, when in lunar orbit two of the astronauts enter the LM, it separates from the CSM and lands on the moon using the Descent Propulsion System (DPS). Following astronaut EVA activities, the LM launches from the moon using the Ascent Propulsion System (APS).

The LM ascent stage contains the crew compartment, APS, RCS thrusters for attitude and translation control, and LM GNC systems, the PGNCS, Abort Guidance System (AGS), and the Rendezvous Radar (RR).

The LM descent stage with its Landing Radar, DPS and propellant tanks are left on the moon, providing a launch pad for the ascent stage.

After rendezvous with the CSM and LM closeout/separation, the now useless ascent stage is directed by Mission Control to impact the moon's surface for seismology studies. For Apollo 10 the LM was jettisoned into solar orbit where it remains today.

A NOTE ON SPACECRAFT ORBITS

It's all about gravity and velocity – with imagination

A short, mostly experiential understanding - for people in a hurry

Satellite orbits Imagine the earth to be a perfect sphere with a radius of 4,000 miles (this is very nearly true), not rotating and without an atmosphere. Imagine if you were to drop a pebble or a piano from 4 feet (or a spacecraft from 100 miles) above this perfect earth - they would of course all fall straight down – towards the center of the earth – due to gravity.

We also know that if any of these objects, the pebble, piano or spacecraft, were dropped at the same time from the same height, they would fall together and all hit the ground at the same time – **even though their weights were different.**

This is not intuitive, but was proven long ago by Galileo (1564-1642) in his leaning tower of Pisa experiment. It was also proven by skeptic astronaut David Scott, who, while on the moon, held a feather in one hand and a geological hammer in the other, then dropped both. They fell together and hit the lunar surface at the same time.

Now imagine that you "throw" the pebble or piano (or spacecraft) in a direction parallel to the ground – that is, add an impulsive horizontal velocity (a ΔV) to the object. After you let go it would still fall toward the earth due to gravity, but in a curved path and eventually hit the ground some distance in front of you.

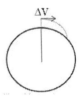

The distance where it landed is dependent only on how much velocity you added, but not dependent on the object's weight – per the Galileo and astronaut Scott experiments.

And, if you threw from the same height the pebble, piano or spacecraft fast enough (about 17,000 miles-per-hour for a circular orbit), it would continue falling toward the earth, but never hit it - the pebble, piano or pacecraft would be "orbiting" around the earth. Of course launching a piano into space takes a much more powerful rocket than launching a pebble.

What is Gravity? Gravity is an *attractive force* that exists between *all* objects in the universe. Here "all" really means all - every sun, star, planet, moon, a person on earth or on the moon, a feather or a rock, etc. – there is an attractive force between all objects. It's sort of like magnetism but it's not because it exists between *everything* in the universe - the "pull" each object has on all the other objects, and vice versa.

Newton's law of universal gravitation, published in 1687, states that "any two bodies in the universe attract each other with a force (F) that is directly proportional to the product of their masses (m_1 and m_2) and inversely proportional to the square of the distance between them (r).

In equation form, this is: $F = G \times (m_1 \times m_2) / r^2$. G is the gravitational constant, which remains the same wherever it is applied in the Universe.

This interesting equation says that the more mass an object has, the larger its attractive force is on the other objects, but the farther away an object is from any other object, the attractive force between them is **reduced by a whole lot - by the inverse square** of the distance they are apart.

This is very good because it simplifies orbit prediction calculations when high precision is not needed. An Apollo example - when in lunar orbit, the LM PGNCS used simple two-body Kepler orbit prediction to calculate the TPI and MCC rendezvous burns because this method was almost as accurate as the more precise but time-consuming numerical integration methods.

Also as a consequence of this, Apollo used two separate coordinate systems for keeping track of LM and CSM state vectors (their position and velocity at a point in time). One coordinate system had its origin at the earth's center of mass, the other at the moon's center of mass. The coordinate system was shifted from earth-centered to moon-centered when the spacecraft's position was less than 40,000 miles from the moon, then shifted back from moon-centered to earth-centered when the spacecraft's position was greater than 40,000 miles from the moon.

Similarly, depending on the precision needed one can ignore the sun's gravitational pull in calculating the orbit of the moon around the earth or in calculating a satellite's orbit around the earth or moon because even though the sun's mass is huge, it is about 93 million miles distant from the earth/moon, and therefore has only a small effect on satellites orbiting around them.

Likewise the moon's gravitational pull could be ignored for satellites orbiting the earth, or for when the earth is orbiting around the sun. Kepler predictions work well for all these.

Momentum Newton's first law of motion is about momentum: "*an object either remains at rest or continues at a constant velocity unless acted upon by a force.*" The force for example could be gravity pulling on an object, or an object's rocket engine that pushes on it.

For an orbiting satellite, a constant tug-of-war takes place between the satellite's tendency to move in a straight line, or momentum, and the tug of gravity, pulling it towards the earth.

In our solar system, where the planets revolve around the sun, and moons orbit their planets, etc., all are moving in inertial space, but it's momentum and the large distances they are apart that keeps them from crashing into each other due to gravitational attraction - think about it.

Orbits Shown here are various orbits relative to the earth (or moon) dependent on their velocity when placed into orbit (at different altitudes for the diffefrent orbits shown). "Throwing" the object faster and faster would result in increasing values of e (orbit eccentricity), from a circular orbit where e=0 (the grey circle) to a parabolic or "escape" trajectory (i.e., would never return to the earth or moon) with e=1, to a hyperbolic trajectory with e>1 (for the simple two-body case where the smaller body would just pass on by the earth or moon).

EARTH OR MOON

TLI placed the Apollo 11 spacecraft into an "escape" trajectory towards the moon. But as it approached the moon, relative to the moon's gravitational attraction it was on a hyperbolic trajectory, requiring the LOI burn to be "retrograde" – opposite its direction of motion - to reduce its velocity consistent with a circular orbit around the moon.

Where did the 17,000 mph orbital velocity come from?

It takes about 1.5 hours for a low earth-orbiting satellite to make one revolution around the earth, and given that the earth's circumference is 25,000 miles, the velocity needed for our rock (or anything) to travel completely around the earth is about 25,000 miles divided by 1.5 hours or 17,000 mph.

One can do a similar calculation for lunar orbits. The period of a low altitude object orbiting the moon is about 2 hours - for one complete revolution. Given the moon's radius is 1,080 miles, we find a lunar orbiting spacecraft zipping above the moon's surface at about 3,700 mph.

Earth's Rotation

The earth rotates in inertial space at one revolution every 24 hours, west to east, so all points on the equator are moving about 1,000 mph in an easterly direction (25,000 miles circumference divided by 24 hours). That is why satellites are typically launched in an easterly direction, to take advantage of the earth's rotation in attaining the velocity needed to get into orbit.

Cape Canaveral, because it's at 28.6 degrees latitude north of the equator, moves east at a slower velocity, about 900 mph, because locations north or south of the equator have a shorter distance to go in 24 hours than at the equator, therefore move slower in inertial space.

This rotational velocity difference due to latitude is also why cyclones, hurricanes, etc. spin counterclockwise in northern latitudes and clockwise in southern latitudes.

APOLLO - MISSION TRAJECTORIES

(Apollo 17 trajectory shown)

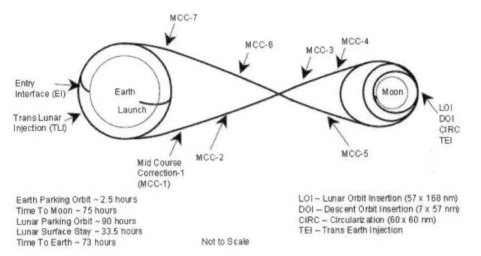

Earth Parking Orbit ~ 2.5 hours
Time To Moon ~ 75 hours
Lunar Parking Orbit ~ 90 hours
Lunar Surface Stay ~ 33.5 hours
Time To Earth ~ 73 hours

Not to Scale

LOI – Lunar Orbit Insertion (57 x 168 nm)
DOI – Descent Orbit Insertion (7 x 57 nm)
CIRC – Circularization (60 x 60 nm)
TEI – Trans Earth Injection

After TLI by the Saturn S-IVB, MCC-H provides the ΔV targeting parameters to the CSM PGNCS for SPS execution of the LOI, DOI, CIRC and TEI propulsive burns shown here. Nominally zero MCC maneuvers are executed as needed to correct for trajectory perturbations.

Free return and hybrid trajectories. The TLI burn for the first three lunar missions (Apollo 8, 10 & 11) placed the CSM/LM spacecraft onto a "free-return" trajectory wherein:

- The spacecraft would pass behind the moon as needed to accomplish the Lunar Orbit Insertion (LOI) burn. LOI is a retrograde burn (slowing down the spacecraft velocity) that places the CSM/LM spacecraft into lunar orbit.

- If the LOI burn was not done (due to a system failure), the spacecraft would be on a return path suitable for reentry to the atmospheric entry corridor.

Starting with Apollo 12, a hybrid trajectory was used. The TLI burn was targeted to place the spacecraft into a highly elliptical Earth orbit that fell short of the moon, but was on a free return path to the atmospheric entry corridor.

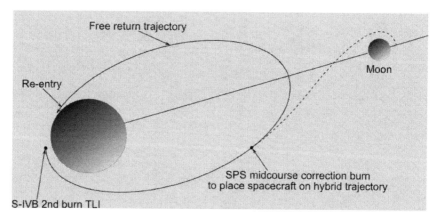

After activation and checkout of the LM, a mid-course maneuver was performed about 30 hours into the mission to change to a trajectory that was not free return but targeted to optimize reaching the desired landing site.

This trajectory retained the safety characteristics of initially being on a free return path and only departed from that once all LM systems were checked out and found operational - the LM then able to provide backup propulsion and other capability later if needed.

LUNAR RENDEZVOUS TRAJECTORIES

The Gemini program had developed and tested in earth orbit methods used to rendezvous the LM with the CSM "mother ship" in lunar orbit. The Apollo missions initially used the four impulse coelliptic maneuver sequence, done after LM launch into a safe lunar orbit:

1. CSI ΔV about 30 minutes after LM safe-orbit insertion.
   (typically 30 by 10 n. miles)

2. CDH ΔV at the apolune after CSI
   - so LM orbit is a constant 15 n. miles below CSM's orbit.

3. TPI maneuver targeted for a CSM intercept - about 43 minutes later

4. Terminal braking to effect the rendezvous

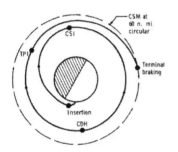
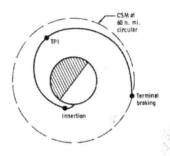

COELLIPTIC SEQUENCE RENDEZVOUS	**SHORT RENDEZVOUS SEQUENCE**

The coelliptic sequence was used for the first two lunar landing missions, Apollo 11 and 12, and planned for Apollo 13. The two impulse short rendezvous technique was used for the remaining four missions - Apollo 14, 15, 16 and 17, done primarily because it eliminated one lunar orbit in the sequence, shortening the rendezvous day by 2 hours.

However the coelliptic sequence targeting capability was retained as an onboard backup capability. This was decided after extensive dispersion analyses and analysis of many contingency situations, including various descent abort and launch trajectories as well as CSM rescue of a disabled LM.

ONBOARD & GROUND GNC SYSTEMS

PGNCS   <u>Primary Guidance, Navigation and Control System</u> pronounced "pings" both the CM and LM contained a PGNCS, an Inertial Guidance System. Major components are the IMU, AGC, DSKY and Software written by MIT.

Despite the word "primary" in its name, PGNCS data was not the main source of navigation information. This was provided by tracking data from NASA's Manned Space Flight Network (MSFN) and processed by the computer system at Mission Control. The position and velocity estimates (state vectors) that resulted were more accurate than those produced by PGNCS - using optical sightings.

As a result, the astronauts were periodically given state vector updates to enter into the AGC.

GNCS was still essential to maintain spacecraft orientation and accomplish steering control during maneuvering burns, including lunar landing and take-off, and as the prime source of navigation data during planned and unexpected communications outages. PGNCS also provided a check on ground data.

IMU   **Inertial Measurement Unit**. The IMU consisted of a gimbaled **stable** member with three gyroscopes and three accelerometers mounted in it. Feedback loops used signals from the gyroscopes to control motors at each axis to keep the stable member fixed with respect to inertial space.

Inertial guidance systems are not perfect. The Apollo IMU's drifted about one milliradian per hour, thus it was necessary to "realign" the inertial platforms periodically by sighting on known stars with the CM and LM optical systems.

AGC   **Apollo Guidance Computer**. The AGC had a 16-bit word length, with 36,864 words of ROM and 2,048 words of RAM. Astronauts communicated with the AGC using a numeric display and keypad called the DSKY.

REFSMMAT "**Reference to Stable Member Matrix**" It was necessary to express the spacecraft's position and velocity relative to the Earth and Moon in a variety of coordinate systems. REFSMMAT was the transformation matrix between an earth-centered (or moon-centered) inertial (fixed relative to the stars) coordinate frame and the PGNCS IMU stable member coordinate frame.

Different REFSMMATS would be calculated by the MCC-H flight controllers for major engine burns or for use during a particular mission phase; e.g. Launch from the earth or moon, the LOI burn, entry, etc. After the REFSMMAT was uplinked to the spacecraft, the IMU would be realigned to that spatial orientation. The calculated REFSMMAT defined an IMU's orientation such that the crew's FDAI (8-ball) display, which showed Pitch, Yaw and Roll angles of the spacecraft attitude about the IMU such that is was easy for the crew to monitor the spacecraft's attitude during that burn or mission phase.

It was also calculated such that the IMU would not go into gimbal lock for the expected attitude changes throughout the burn or upcoming maneuver.

For the Apollo 17 mission, eight different REFSMMAT's or IMU alignments were used.

STATE VECTOR __A State Vector__ consists of six numbers, three components of position and three components of velocity, that define for a specific point in time the spacecraft's position and velocity in a Basic Reference Coordinate system (BRC).

The BRC is an orthogonal inertial coordinate system whose origin is either at the earth or moon center of mass. The coordinate system is shifted from earth-centered to moon-centered when the spacecraft's position is less than 40,000 miles from the moon, and likewise shifted from moon-centered to earth centered when the spacecraft is greater than 40,000 miles from the moon.

All navigation stars, lunar ephemerides and vehicle state vectors are referenced to the BRC system.

RTCC __Real-Time Computer Complex__ The RTCC was the IBM computing and data processing system used to collect, process and send to the Apollo Mission Control Center-Houston the information needed to direct all phases of the Apollo missions. It computed spacecraft trajectories using DSN tracking data and determined the different ΔV maneuvers needed, as well as REFSMATT information for the spacecraft.

RTCC worked in real-time – very fast to minimize the time between receiving external data and solving a computing problem.

IBM 7094 computers were used in the RTCC during the Gemini program and on the first three Apollo missions. Later, IBM System/360 mainframes were employed. Two computers were used during a mission: one was primary; the other operated identically as standby.

"ROCKET SCIENTIST" One term I try not to use is "Rocket Science" or "Rocket Scientist." However, it is a much used term due to people not considering the distinction between science and engineering.

I like to think of it as a spectrum – on one end is basic science, which aims to understand of the laws of nature without regard for their practical application; at the other end is engineering, which aims to create technological devices to shape our world to human purposes.

Neil Armstrong, the first person to step onto the moon, explains in his YouTube video this difference simply as: "Science is about ***what is***; Engineering is about ***what can be***."[2]

A good example: Wernher von Braun and his team that built the huge Saturn V launch vehicle were "***rocket engineers***." Although von Braun earned a doctorate in Physics in 1934, he never worked a day in his life as a scientist - he was an engineer and a manager of engineers, and used that vocabulary when talking with his professional peers.

---

[2] Neil Armstrong's "on Being a Nerd" -
http://www.youtube.com/watch?v=6aQlbqHZuYk.

TRW's ROLE

TRW Inc. had three major activities supporting the Apollo Program:

1. **The LM Descent Engine.**

    TRW designed and built the descent engine for the Apollo Lunar Landing Module (LM) under subcontract to GAEC, the first throttleable rocket engine for manned space flight. This engine performed as designed for all six of the successful lunar landing missions, and was also used for the three critical burns needed to place the crippled Apollo 13 spacecraft onto a safe return-to-earth trajectory, providing an effective backup to the damaged Service Module and its Propulsion System.

    The first DPS burn was accomplished about two hours after the oxygen tank explosion, a 35-second burn to place the crippled spacecraft back onto a free return trajectory. The next burn was a long 5-minute Transearth Injection burn intended to shorten the return to earth time, and the last a short 14-second midcourse correction burn.

2. **The LM Abort Guidance System**

    TRW designed and built, also under subcontract to GAEC, the Abort Guidance System (AGS) for the LM, a backup system used to control a LM abort from descent to the moon if the Primary Guidance, Navigation and Control System (PGNCS) should fail.

    There were reasonable concerns about the possibility of the LM PGNCS failing while the LM was descending onto the rocky surface of an unknown and hostile world. As the PGNCS and its controlled propulsion systems were complex, new and never used before, it was decided that if the PGNCS did fail, there should be a simple and reliable way to "get the hell out of there." AGS was the solution. It could take over at any time and guide the LM first into a safe lunar orbit and then calculate and control the APS and RCS burns needed for rendezvous back with the CSM "mother ship." AGS could not be used, however, for continuing with the landing if the PGNCS should fail.

    AGS was the first inertial navigation system to use a strap-down Inertial Measurement Unit rather than a gimbaled gyro-stabilized IMU as used by PGNCS - much lighter and smaller in size than the PGNCS, although inherently less accurate and more prone to gyro drift and accelerometer errors, but adequate for its intended purpose.

The LM PGNCS performed well for all Apollo flights with a LM; therefore AGS as a PGNCS backup was never needed. However, for Apollo 13, after the TEI burn AGS was used instead of the PGNCS for DPS/RCS burns and attitude control over the next 2-plus days - to minimize the usage of electrical power and cooling water.

3. **TRW Houston Operations.**

Across the street from the Manned Spacecraft Center (MSC), TRW Houston Operations had over 900 employees with approximately 600 professional engineers, computer programmers, analysts, mathematicians and scientists that provided a very significant capability in quantity and depth in orbital mechanics, trajectory design and analysis, targeting, simulation, and systems engineering services in support of the NASA/MSC Apollo Spacecraft Program Office (ASPO).

NASA's decision to build up a TRW capability in Houston had its roots in the Air Force ballistic missile programs. STL-TRW had a comprehensive systems engineering role with respect to the AF ballistic missile programs, and the Mercury and Gemini programs both used AF Atlas and Titan II boosters as their launch vehicles.

The utilization of the guidance system on the Atlas, and the close interaction between the software in the Atlas guidance system and the orbital aspects of the Mercury flight, brought TRW engineers into close contact with key members of the NASA Flight Operations Directorate (FOD), which led to the decision to build up a Houston capability in support of that directorate. This was TRW's Mission Trajectory Control Program (MTCP), created to specifically support the Mission Planning and Analysis Division within the FOD - doing flight planning, mission planning and real-time support.

The second major piece of TRW Houston's contract was entitled Apollo Spacecraft Systems Analysis Project (ASSAP), which came about due to Dr. Shea's conviction that the systems analysis and systems engineering capability as applied to complex programs (and convincingly demonstrated by TRW during the course of the ballistic missile programs), was needed in direct support of MSC. Before joining NASA, Dr. Shea had been active in the AF ballistic missile programs at Bell Labs, AC Spark Plug and TRW.

It was under this contract that TRW identified the need for Apollo Mission Techniques, initiated their development, and played a key role in their ongoing development – a significant and wide-ranging effort that contributed in a major way to the success of the Apollo moon landing program.

Part II of this book includes an overview of the Apollo Mission Techniques activity.

Part III and the appendices describe that activity more broadly and in greater detail.

WHO ACCOMPLISHED THIS FEAT?

Days before launch of Apollo 11, *Time* magazine published the article "WHO MADE IT POSSIBLE," identifying those men they believed "eliminated obstacles that might have delayed the program indefinitely." From that article:[3]

> ➢ Dr. John C. Houbolt, 50, former chief of theoretical mechanics at NASA's Langley Research Laboratories in Hampton, Va. Houbolt, a civil engineer, is responsible for the lunar-orbit rendezvous that is the key maneuver in Apollo's entire flight plan. In what he remembers as "an intuitive flash," Houbolt realized that tremendous weight savings would be gained by this rendezvous method, permitting the use of a smaller launch vehicle. Houbolt fought a two year battle, finally putting his job on the line by appealing directly to NASA headquarters. His arguments prevailed in the fall of 1962."

> ➢ Dr. Wernher von Braun, 57, director of the Marshall Spaceflight Center in Huntsville, Ala. Transported to the U.S. by American intelligence officials in 1945, along with 126 other German scientists who had been working on the V-2 rocket at the Baltic base of Peenemunde, Von Braun has directed development of rocket-launch vehicles from the earliest Redstone. Von Braun helped develop the ablative heat shield, which dissipates the searing heat of re-entry by flaking off in harmless fiery pieces. Von Braun, perhaps more than any other man, has been the driving force behind the moon program.

> ➢ Dr. Charles Stark Draper, 67, director of the Instrumentation Laboratory at the Massachusetts Institute of Technology. To solve the problems of navigation, NASA went straight to the nation's leading authority on inertial guidance. The system devised by Draper for Apollo includes telescopes, a sextant, and a computerized inertial reference "platform" that tells astronauts where they are in space, where they are headed and how fast. But how could they be sure that it would work?, the NASA brass wanted to know. "I told them I'd go along and run it myself," recalls Draper. The onboard navigation systems have proved so accurate that, if they had to, the crew of Columbia could fly to the moon and back without help from ground controllers.

---

[3] *Time* magazine Special Supplement "TO THE MOON" dated July 18, 1969; page 29.

Other men were almost as indispensable. Maxime A. Faget, director of engineering and development at Houston's Manned Spaceflight Center, designed Apollo's command and service module. Dr. George E. Mueller, NASA' top official for manned spaceflight, introduced a time-saving technique known as "all-up testing," in which all three rocket stages are tested together. Christopher Kraft, director of flight operations since 1961, and George Low, manager of the Apollo program, brought a sense of cool discipline to the nerve-racking operations in Houston.

Then too, there is Donald K. ("Deke") Slayton who selects and trains the astronauts. The professionalism of the Apollo Crews is a reflection of Slayton's success.

[end of *TIME* magazine article]

George E. Mueller and Samuel C. Phillips

In my opinion, George Mueller (pronounced "Miller") is the single individual most responsible for the success of the Apollo program, followed closely by Samuel Phillips. The program management controls they brought from their Minuteman ICBM experience to the Apollo program, their tireless efforts at effective implementation of these at MSC, KSC *and especially at MSFC,* and Mueller's promulgation of "all-up testing" provided major cost benefits to the program - as well as meeting Kennedy's moon landing schedule goal.

Who knows, without their resolute implementation of these changes over their 5-plus years of directing the moon landing program, the schedule lengthening and consequent increasing costs (the "marching army" syndrome) might very well have eventually led to program cancellation.

Here is more about their considerable contributions to Apollo's success:

George E. Mueller

Dr. Mueller joined NASA as head of the Office of Manned Space Flight (OMSF) on September 3, 1963. Prior to that, NASA's Administrator James E. Webb had sounded him out for this job, but he would only agree if the directors of MSC, MSFC, and KSC were to report directly to him as head of the OMSF. This was done and Mueller accepted, although he took a substantial pay cut. This reorganization of NASA and the OMSF was announced in November, 1963, and the three centers responsible for launch vehicle, spacecraft and launch site development were all his direct responsibility.

Mueller had joined Ramo-Wooldridge (the RW part of TRW) in 1954, became director of the Electronics Laboratories in 1957 and progressed over the years to Vice President, Research and Development for Space Technology Laboratories (STL). TRW/STL had been the Systems Engineering and Technical Direction (SETD) contractor for the Minuteman ballistic missile program – a major program for TRW since 1954 that stretched over many years.

While at STL Mueller had become familiar with program management methods used for the Air Force ballistic missile programs and had concluded that at OMSF there "wasn't any management system in existence" - no formal means existed to determine and control hardware configurations and therefore no way to determine realistic costs or schedules. He would have to "teach people what was involved in doing "program control."

His answer to this was to implement management methods as developed in the Air Force and at TRW for the ICBM program.

These were more rigorous, paperwork-heavy mechanisms for program control over costs and schedule. He centralized Apollo Management in a headquarters office that would communicate - on a daily basis directly with corresponding offices at MSC, MSFC and KSC as shown here.

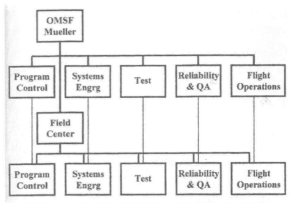

Mueller's organization at OMSF had direct interfaces with corresponding organizations at MSC, MSFC and KSC field centers.

Systems Engineering Management – based upon my TRW project experience this is a methodology and process used to ensure that large, technically complex systems can be successfully developed meeting cost, schedule and performance requirements. These systems typically contain sophisticated hardware & software, with subsystems developed by different companies, and all technical and management elements must work smoothly together for project success.

This is the stuff of configuration management with CCB's controlling a documented baseline configuration, specifications that define functional, performance, interface and test requirements, formal ICD's to define and control interfaces between elements provided by different organizations, a WBS with work package managers responsible for cost, schedule and performance of their work units, project schedules with PERT charts/analysis showing critical path dependencies, structured design reviews, test plans and procedures, etc., etc. Much formal documentation to ensure accurate assessment of project progress and identified problems are understood early by management levels that are able to resolve them in a timely fashion.

The recognized "bible" for this is *Air Force Systems Command Manual 375-5, System Engineering Management Procedures.* Major TRW projects I have worked on have used most, if not all, elements of this methodology.

Mueller developed effective relationships with the presidents of Apollo's main contractors as well - NAA for CSM and the Saturn V's second stage, GAEC for the LM, etc. similar to what was done on the US Air Force Minuteman program. It was his influence on their organization's structure and operations (establishing primacy of their projects by implementing effective CCB's, etc.) that led to timely and effective problem resolution and maintaining tight program schedules and therefore costs.

Almost everyone who worked with Mueller on Apollo agreed he was technically brilliant and exceedingly capable. One of his colleagues described working for him as a "piece of cake.........the epitome of politeness, but you know down deep he's just as hard as steel!"

Samuel C. Phillips

Mueller found that he could not always find the right people with appropriate program management skills. Again, using his background with Air Force projects, he sought Webb's permission to bring in skilled Air Force program managers and proposed Minuteman program director Colonel Samuel C. Phillips as Apollo program director in OMSF. Webb agreed, and so did AFSC chief General Bernard Schriever. Phillips in turn agreed and brought with him 42 mid-grade officers and eventually 124 more junior officers.

While they had not worked directly together, Mueller knew he wanted Phillips on his team. Phillips had hoped to return to SAC for another operational tour and he was reluctant to join NASA. Nonetheless, he stepped right in and took charge. Though initially assigned as Deputy Apollo Program Director under the dual-hatted Mueller, he acted from day one, Jan. 1, 1964, as Program Director, a title he formally received nine months later.

Phillips brought a range of highly-developed program management disciplines with him from the B-52 and Minuteman programs, including configuration management, change control with associated cost estimating; associate contractor relationships built on mutual trust and frank communication; and a systems engineering approach that produced a "full visibility" baseline configuration.

George Mueller (left) and Sam Phillips (right) NASA photo

Von Braun was at first upset by the many changes introduced by Mueller and Phillips; they seemed to undercut the role of center director and his conservative approach to rocket development. He complained about the size of the engineering staff in Washington . . . "My reply to this whole problem is . . . to cut them down from 400 to 1 and we'll be in good shape." MSC director Gilruth agreed, but was reluctant to go along with von Braun in cooking up a conspiracy against headquarters. Webb noted that the MSFC director was "very impatient of paperwork – per von Braun "we can lick gravity, but sometimes the paperwork is overwhelming."

A serious problem at MFSC was technical staff in the labs (Guidance and Control, etc. whose directors worked directly for von Braun) giving "technical direction" (engineering changes) to contractors while ignoring "normal contractual procedures." This problem was worked quietly and effectively by Mueller's team such that, eventually, von Braun gave the labs clear and strong direction that "their role was strictly one of support" to the contract managers, who controlled the 90 % of MSFC's budget that went to contractors.

Ultimately von Braun realized the overall benefits of their methods, and even when necessary to consolidate a consensus among the agency leadership he would overrule his own people. Phillips, when later asked about von Braun being a good team player, made the interesting observation that he *became* a good team player.

After the Apollo 1 fire January 27, 1967, NASA Administrator James Webb became distrustful of Mueller, but commented, "even if I wanted to, I couldn't fire him because he was manager of our successful Apollo project, and one of the ablest men in the world ... The last thing I wanted was to lose him, but I also had another desire, which was not to let his way of working create too many difficulties."

To address Apollo's slipping schedule and huge cost overruns, Mueller promoted the "all-up testing" approach used successfully on both USAF's Titan II and Minuteman programs. He emphasized clearly that all-up meant "simultaneous" or "all at once" - the complete launch vehicle *and* spacecraft for all development flights would be functional; "live" and as close to the lunar configuration as possible, and beginning with the very first test flight.

This approach at first was not acceptable to von Braun and his team of conservative rocket engineers. Their test plan called for the first live test to use the Saturn's first stage with dummy upper stages. If the first stage worked correctly then the first two stages would then be live with a dummy third stage and so on, with at least ten test flights before a manned version was put into low earth orbit.

In time Dr. Mueller proved persuasive enough to overcome von Braun's reservations, and it was "all up" for the mammoth Saturn V test program. As von Braun stated, "It sounded reckless, but George Mueller's reasoning was impeccable" – to wit – "Water ballast in lieu of a second and third stage would require much less tank volume than liquid-hydrogen-fueled stages, so that a rocket tested with only a live first stage would be much shorter than the final configuration. Its aerodynamic shape and its body dynamics would thus not be representative. Filling the ballast tanks with liquid hydrogen? Fine, but then why not burn it as a bonus experiment? And so the arguments went on until George in the end prevailed."

The very first test flight of the giant Saturn V was on November 6, 1967 in an all-up configuration. The three Saturn V stages were "live" with a live (but unmanned) payload, the CSM. It flew a sophisticated trajectory - after earth orbit insertion by the S-IVB, a second SIVB burn, emulating the translunar injection burn, and then two burns by the CSM's Service Propulsion System would place the CSM on a trajectory that reenters the atmosphere under conditions simulating a return from the Moon - for a heat shield test. This was done, and this complex first flight of the unmanned Saturn V, Apollo 4/SA-501 was a complete success.

Mueller's concept of all up testing worked. The first two unmanned flights of the Saturn V were successful, the second less so due to the "pogo" problem (pages 105-108 has more on this problem and the Apollo 8 launch decision), but the third Saturn V launch was a complete success and put Frank Borman's Apollo 8 crew into orbit round the Moon. After the successful Apollo 9 and 10 launches, the sixth Saturn V launch carried Neil Armstrong's Apollo 11 crew toward the first moon landing.

In an interview Mueller acknowledged what would have happened if all up testing had failed, "The whole Apollo program and my reputation would have gone down the drain."

THEY WERE ENGINEERS

Apollo 11 liftoff L-R Mathews, von Braun, Mueller and Phillips

Charles W. Mathews - graduated from Rensselaer Polytechnic Institute in Troy, NY as an aeronautical engineer. He worked as an engineer at Langley Research Center in Virginia and at the Manned Spacecraft Center in Texas before joining the Office of Manned Space Flight (OMSF) as Gemini program manager, overseeing the 10 Gemini missions that were flown by most of the astronauts who later traveled to the Moon.

Wernher von Braun – pursued his early interest in rocketry at one of the best engineering schools in Europe, the Berlin-Charlottenburg Institute of Technology (now the Technical University of Berlin). He led a team of German engineers/scientists in developing the V-2 rocket used in World War II. After the war, he was brought to the United States under operation "paperclip" and became the father/leading figure of rocket technology and space science in the United States. His team of German engineers in Huntsville launched the first American satellite and developed the huge booster rockets used for all Apollo moon flights.

George E. Mueller – earned a BSEE degree from the Missouri School of Mines and MSEE degree from Purdue University. He later taught electrical engineering and the new field of system engineering at Ohio State University while completing his Ph.D. in Physics. By the mid-1950s he was consulting with major aerospace companies. Joining Ramo-Wooldridge in 1957 as director of the Electronics Laboratories, he quickly rose to management positions at TRW/STL, and was Vice President, Research and Development, when NASA hired him in 1963 to lead the OMSF.

General Samuel C. Phillips - earned a BSEE degree from the University of Wyoming and upon graduation in 1942 entered the Army and transferred to the air component. As a young pilot, he served with distinction during World War II, earning combat honors. Later he earned the MSEE degree from the University of Michigan in 1950. In the Air Force, he rose to Director of the Minuteman ICBM Program, then at Mueller's request he was loaned to NASA as Director of the Apollo Program within the OMSF.

Apollo Program Supporting Organizations

INTRODUCTION

At the height of the Apollo program, NASA had 35,000 employees and more than 400,000 contractor employees in thousands of companies and universities across the U.S.

The four principal organizations within NASA responsible for successfully accomplishing the Apollo moon landing program were:

.NASA Headquarters – provided overall guidance and direction for the program including assignment of roles and responsibilities for the various NASA centers and arbitrating among them.

James E. Webb ran NASA from the beginning of the Kennedy administration through the end of the Johnson administration, overseeing all Mercury and Gemini orbital flights and dealing with the Apollo 1 fire. He resigned October 7, 1968, four days before launch of the first manned Apollo flight, Apollo 7. He was succeeded by Dr. Thomas O. Paine.[4]

Dr. George E. Mueller headed the OMSF first under Webb, then Paine. Under Dr. Mueller were the Gemini and Apollo Program Directors, Charles W. Mathews and General Samuel C. Phillips, and the three field center directors; Robert Gilruth, (MSC), Wernher von Braun (MSFC), and Kurt Debus (KSC).

The Manned Spacecraft Center (MSC) in Houston, Texas (now renamed the Johnson Space Center - JSC). MSC was responsible for managing the development of the two Apollo spacecraft - the Command/Service Module (CSM) and the Lunar Module (LM). MSC was also responsible for Astronaut training and real-time flight control of the Apollo spacecraft on its 7-10 day mission to the moon and back – via the Mission Control Center-Houston (MCC-H).

Prime Spacecraft contractors managed by MSC were North American Aviation (NAA) for the CSM and Grumman Aerospace Engineering Company (GAEC) for the LM. IBM was responsible to MSC for the MCC-H computers

---

[4] James Webb was a Democrat with close ties to President Johnson. When Webb learned Johnson chose not to run for reelection, he resigned to allow the next president, Richard Nixon, to choose Paine as his own administrator. Paine ran NASA until Sept 15, 1970.

and software, and MIT did the hardware design and software development for the Apollo Guidance Computers in the CSM and LM.

The Marshall Space Flight Center (MSFC) - Wernher von Braun's team in Huntsville, Alabama, were the rocket engineers – responsible for building the giant Saturn V rocket (pronounced "Saturn five") which launched the Apollo Spacecraft onto a trajectory towards the moon.

Prime Saturn V contractors managed by MSFC were Boeing for the S-IC first stage, North American Aviation (NAA) for the S-II second stage, Douglas Aircraft Company for the S-IVB third stage, and IBM for the Instrument Unit containing the Saturn V Guidance & Control system.

NAA's Rocketdyne division built all of Saturn V's rocket engines - the five powerful F-1 engines for the S-1C first stage; the five J-2 Liquid Hydrogen-Liquid Oxygen engines for the S-II second stage; and the single J-2 engine for the S-IVB third stage.

As of today, early 2019, the Saturn V remains the tallest, heaviest, and most powerful (highest total impulse) rocket ever brought to operational status, and holds records for the heaviest payload launched and largest payload capacity to low Earth orbit of 310,000 lb., which included the third stage and unburned propellant needed to send the Apollo CSM and LM to the moon.

It is worth noting that the Saturn V was successfully launched 13 times from KSC with no loss of crew or payload. There were 12 Apollo mission launches and a last Saturn V launch for the Skylab space station.

Kennedy Space Center (KSC) - NASA's Saturn V launch facilities in Florida. Kurt Debus was responsible for KSC design, development and operations throughout the Apollo program.

OTHER SUPPORTING ORGANIZATIONS

Jet Propulsion Laboratory (JPL) NASA's *JPL* in Pasadena, California, is the leading U.S. center for the robotic exploration of the solar system and *a* division of *Caltech* - the California Institute of Technology. JPL's intellectual environment and identity are profoundly shaped by its role as part of Caltech (just 7 miles away) and its long history of leaders drawn from that university's faculty.

JPL provided support to the Apollo program in two critical areas:

(1) **Management and operation of NASA's Deep Space Network.** The DSN is a worldwide system for spacecraft communications providing for the acquisition, transmission, processing, display, and control of spacecraft tracking and communications information used for spacecraft and mission control, spacecraft performance monitoring, scientific measurements and photography.

(2) **Management and operation of the Surveyor program.** The primary objective of this program was to determine if the moon's crust would support a soft landing, answering questions such as how deep was the dust on the Moon. The Surveyor missions provided photographs and measurements of the composition and surface-bearing strength of the lunar crust, and readings on the thermal and radar reflectivity of the soil.

Between May 30, 1966 and January 7, 1968, seven Hughes Aircraft Company's Surveyor spacecraft were injected by Atlas-Centaur launch vehicles onto a trajectory targeted directly at the landing site. Near the moon, retrorockets slowly decelerated the spacecraft from its impact trajectory to accomplish the soft landing.

Surveyor 1 on May 30, 1966, was the first US spacecraft to soft-land on the moon. The Soviets Luna-9 spacecraft had previously soft-landed on the moon just 4 months earlier, on February 3, 1966.

Two of the seven Surveyor spacecraft crashed onto the moon, the five others successfully soft-landed, supporting Apollo and gathering other scientific data.

Langley Research Center – The Boeing Lunar Orbiter Project

Although JPL's primary mission within NASA was the leading center for development and operation of unmanned spacecraft, the NASA Langley Research Center in Hampton, Virginia, was given the lunar orbiter project for three fundamental reasons:[5]

1) JPL's manpower and management capabilities at that time were stretched thin by the Ranger and Surveyor projects,

2) Langley had proven itself to be very successful in project management,

3) wider distribution of operational programs among NASA field centers would allow the centers to develop new and varied capabilities for future NASA ventures.

Langley managed Boeing's development and conduct of five Lunar Orbiter missions between 1966 through 1967 to map the lunar surface before the Apollo landings. A powerful Eastman-Kodak camera and other instruments provided detailed mapping of the lunar surface, determination of the Moon's size, shape and surface topography, its gravitational and magnetic fields. Meteoroid detection and radiation measurements were also done.

On Aug. 14, 1966, Lunar Orbiter-1 was the first U.S. spacecraft to orbit the moon (the Soviets had earlier launched their Luna 10 Orbiter on March 31, 1966).

All five missions were successful, and 99% of the Moon was photographed with a resolution of 60 meters or better. The first three missions, flown at low inclination orbits, were dedicated to imaging 20 potential lunar landing sites. The fourth and fifth missions were devoted to broader scientific objectives and were flown in high altitude polar orbits.

---

[5] "Destination Moon: A History of the Lunar Orbiter Program" - Chapter II. Langley enters the picture. https://history.nasa.gov/TM-3487/contents.htm

THE SOVIETS IN THE SPACE RACE

During the time of the space race, much of the information on the rocket and space programs of the Soviet Union was classified, considered a state secret. And for propaganda value announcement of mission outcomes was sometimes delayed until success was certain, and failures were often kept secret.

As a result of Mikhail Gorbachev's policy of *glasnost* in the late 1980s and dissolution of the Soviet Union in 1991, many previously unknown facts about their space program eventually became known – examples being the crash of the Soyuz 1 spacecraft wherein Cosmonaut Vladimir Komarov died, and the catastrophic failures of the huge N-1 rocket, the intended launch vehicle for the Soviet's manned lunar landing.

This summary of the Soviet's space program is based upon information from various internet sites maintained by NASA, Wikipedia and others, and the book *Two Sides of the Moon*, by David Scott and Alexei Leonov, each former spacecraft commanders telling their individual interwoven stories of the cold war space race.

1959–76 SOVIET UNMANNED LUNAR PROBES

The Soviet's aggressive Luna program resulted in many firsts as follows:

Luna 1 - **the first man-made object to achieve escape velocity and orbit the sun.** Intended to impact the moon, Luna 1 missed the moon by 3,700 miles. **January 1959**

Luna 2 - **the first spacecraft to reach the surface of the moon.** **September 1959**

Luna 3 - **the first photographs of the far side of the moon** were transmitted to earth by this lunar flyby mission. **October 1959.**

Luna 9 - **the first spacecraft to achieve a soft landing on the Moon** and transmit photographic data to Earth from the surface of another planetary body. After the Sun had risen to a 7 degree elevation, the probe began sending the first of nine images of the surface of the Moon (including five panoramas). **February 1966.**

Luna 4 launched on April 4, 1963 was the first of 5 failed attempts by the Soviets to achieve a lunar soft landing on the moon (2 missed, 3 impacted the moon).

Luna 10 - **the first artificial satellite to orbit the Moon.** Perhaps its most important finding was the first evidence of mass concentrations (called "mascons" - areas of high density below the mare basins that distort lunar orbital trajectories. Their discovery has usually been credited to the American Lunar Orbiter series. **April 1966.**

Lunas 11, 12, 14, 19 and 22 were also Lunar Orbit missions.

Luna 19 was the first of the "advanced" lunar orbiters, providing panoramic images of the mountainous region of the Moon between 30° and 60° south latitude and between 20° and 80° east longitude. Also did extensive studies on the shape and strength of the lunar gravitational field and the locations of the "mascons."

Communications with Luna 19 were terminated in October 1972 after a year of operation and more than 4,000 orbits around the Moon.

Luna 15, launched on July 13, 1969, was the Soviet attempt to return lunar soil back to Earth before the US Apollo 11 mission. This didn't work out as planned. The Luna 15 orbiting spacecraft began its descent for a planned soft landing on July 21, 1969 but crashed onto the moon while Apollo 11 astronauts finished their first moonwalk.

Luna 15 was the first of six attempts to return lunar soil samples to Earth. The others were Luna 16, 18, 20, 23 and 24, three of which were successful:

> Luna 16 returned 101 grams in September 1970
> Luna 20 returned 30 grams in February 1972.
> Luna 24 returned 170 grams in August 1976 – the last of the Luna series.

Luna 17, launched November 10, 1970, continued the many successes in Soviet automated lunar exploration. It carried Lunokhod 1, the first in a series of robot lunar roving vehicles that traveled over 6 miles on the lunar surface. It transmitted over 20,000 TV images and 206 high-resolution panoramas, and used a penetrometer to test the soil's mechanical characteristics at more than 500 locations.

1961-65 – EARLY MANNED EARTH ORBIT PROGRAMS

Sergei Korolev was the lead Soviet rocket engineer and spacecraft designer during the 1950s and 1960s. He directed the Soviet Sputnik, Vostok and Voskhod space programs until his death on January 14, 1966.

The Vostok Program

The Vostok Program was the Soviet's first generation of manned space vehicles, encompassing six manned flights from April 1961 to June 1963.

The Vostok spacecraft consisted of a spherical cabin coated entirely in an ablative material to act as a heat shield as it re-entered the atmosphere. There was a window for the cosmonaut to view the earth and an ejector seat for his return. Beneath the cabin was a propulsion module used for the retrograde ΔV maneuver needed to initiate reentry.

Upon reentry, the cosmonaut would eject at about 23,000 feet and descend via parachute, while the crew capsule would land separately as its rough landing could seriously injure the cosmonaut.

The Soviets kept the details and true appearance of the Vostok capsule secret until the April 1965 Moscow Economic Exhibition, where it was first displayed without its aerodynamic nose cone concealing the spherical capsule.

<u>Vostok 1</u>

Vostok 1 was launched April 12, 1961. **This was Yuri Gagarin's one-orbit flight, making him the first human in space.**

Only about one month later, on May 25, 1961, President Kennedy announces the US goal to land a man on the moon by the end of the decade. Shortly thereafter the Soviets begin planning toward a manned moon landing.

Vostok 1 capsule at the Energiya Museum outside of Moscow

Vostok 2 - launched August 6, 1961, a longer 1-day manned flight.

Vostok 3 - launched August 11, 1962, a nearly 4-day flight to determine the health effects of longer space missions.

Vostok 4 - launched August 12, 1962, a 3-day manned flight launched into a nearly identical orbit with Vostok 3 to determine how the two cosmonauts' reactions might differ during a series of tests under similar circumstances.

Vostok 5 - launched June 14, 1963. Due to elevated levels of solar flare activity, the planned 8-day mission was reduced to 5 days.

Vostok 6 - launched June 16, 1963, a 3-day mission commanded by **the first woman in space**. Similar to the Vostok 3 and 4 missions, Vostok 6 was launched into a near identical orbit with Vostok 5.

The Voskhod Program

The Voskhod Program was the Soviet's second generation of manned space vehicles, encompassing just two manned flights - one in October 1964 and one in March 1965.

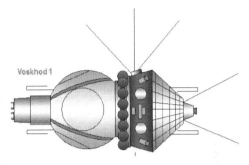

Voskhod 1 launched October 12, 1964 and landed 24 hours later, the first spacecraft flown with a three-cosmonaut crew.

The Soviets advertised another technological achievement during this mission; the first space flight during which cosmonauts performed in a shirt-sleeve-environment. However, this was done because the craft's limited cabin space did not allow room for spacesuits.

Despite the propaganda boasting around Voskhod 1, it was privately referred to by the leadership of the Soviet space program as "a circus" due to the messy process of crew selection, the cosmonauts needing to diet to fit inside the spacecraft, and the very dangerous circumstances of the crew having neither pressure suits nor any way to escape from a malfunctioning launch vehicle.

Voskhod 2 was launched on March 18, 1965, a 1-day 17-orbit mission with two cosmonauts. As part of this mission, **Alexi Leonov became the first person to leave a spacecraft in a specialized spacesuit and conduct a 12-minute "spacewalk."**

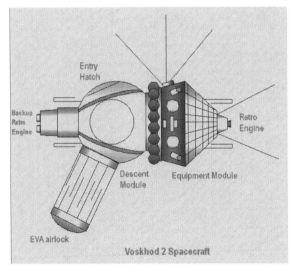

The Voskhod 2 spacecraft had an inflatable airlock allowing EVA while keeping the cabin pressurized so that the capsule's electronics would not overheat (Voskhod electronics used vacuum tubes with heating element; the Gemini spacecraft, also flown in 1965, used solid state electronics).

A fatality was narrowly avoided when Leonov's spacesuit expanded in the vacuum of space, preventing him from re-entering the airlock. In order to overcome this, he partially depressurized his spacesuit to a potentially dangerous level before safely re-entering the ship, but he and Cosmonaut Belyayev faced further challenges when the spacecraft's atmospheric controls flooded the cabin with 45% pure oxygen, which had to be lowered to acceptable levels before re-entry.

The reentry involved two more challenges: an improperly timed retrorocket firing caused the Voskhod 2 to land 240 miles from its designated target area, and the instrument compartment's failure to detach from the descent apparatus caused the spacecraft to become unstable during reentry.

Leonid Brezhnev and a small cadre of high-ranking officials had deposed Khrushchev as Soviet leader on October 14, 1964, one day after Voskhod 1 landed. The new political leaders, along with Korolev, ended the technologically troublesome Voskhod program, cancelling Voskhod 3 and 4, which were in the planning stages, and started concentrating on the race to the Moon.

Voskhod 2 ended up being Korolev's final achievement before his death on January 14, 1966, as it became the last of the many space firsts that demonstrated the USSR's domination in spacecraft technology during the early 1960s. According to historian Asif Siddiqi, Korolev's accomplishments marked "the absolute zenith of the Soviet manned space program."

There was a two-year pause in Soviet manned space flights while Voskhod's replacement, the Soyuz spacecraft, was designed and developed.

1967-74 – THE SOYUZ PROGRAM

The **Soyuz** program, consisting of the Soyuz launch vehicles and Soyuz spacecraft, was the third Soviet human spaceflight program after the Vostok and Voskhod programs.

The Soyuz launch vehicles first introduced in 1966 were based upon the Soviet R-7a ICBM. With subsequent variants they have become the world's most used launch vehicles, reaching a peak of 60 per year in the early 1980s. Despite its age, this rocket family is notable for its low cost and high reliability.

The Soyuz spacecraft were originally designed and developed by the Korolev Design Bureau (OKB-1, now RKK Energia), and variants of that design remain in service today. The spacecraft's basic design was intended to travel to the moon by repeated dockings with upper stages in earth orbit - as evidenced by the mission table list on the next page.

The first Soyuz flight on November 28, 1966 was unmanned.

Soyuz 1, the first manned mission, ended with a crash due to a parachute failure, killing cosmonaut Vladimir Komarov.

Soyuz 3 was launched on October 26, 1968, becoming the program's first successful manned mission. Its mission included docking with the previously unmanned launch of Soyuz 2.

The only other flight to suffer a fatal accident was Soyuz 11 with a crew of three. The cabin depressurized prematurely just before reentry, killing its crew of three - the only humans to date to have died in space.

Beyond 1974

Despite these early failures, Soyuz today is widely considered the world's safest, most cost-effective human spaceflight vehicle, established by its unparalleled length of operational history. Soyuz spacecraft were used to carry cosmonauts to and from Salyut, later to the Mir Soviet space station, and now to the International Space Station (ISS).

Since the retirement of America's Space Shuttle in 2011, all human spaceflights to and from the ISS are now carried out using Soyuz spacecraft and launchers. At least one Soyuz spacecraft is docked to ISS at all times for use as an escape craft in the event of an emergency.

1967-74 – SOYUZ EARTH ORBIT MISSIONS

Mission	Launch	Duration	Landing	Crew	Mission Results
Soyuz 1	**4-23-1967**	**1 day**	**4-24-1967**	**1**	**Cosmonaut died in crash-landing.**
Soyuz 2	10-26-1968	3 days	10-30-1968	0	Unmanned
Soyuz 3	10-26-1968	3 days	10-30-1968	1	Rendezvous with unmanned Soyuz 2 failed docking.
Soyuz 4	01-14-1969	2 days	1-17-1969	3	Docked with Soyuz 5. Crew transfer.
Soyuz 5	1-15-1969	3 days	1-18-1969	3	Docked with Soyuz 4. Crew transfer.
Soyuz 6	10-11-1969	5 days	10-16-1969	2	Joint mission with Soyuz 7 & 8.
Soyuz 7	10-12-1969	5 days	10-17-1969	3	Joint mission with Soyuz 6 & 8.
Soyuz 8	10-13-1969	5 days	10-18-1969	2	Joint mission with Soyuz 6 & 7.
Soyuz 9	6-1-1970	18 days	6-19-1970	2	Cosmonaut endurance test.
Soyuz 10	4-23-1971	2 days	4-25-1971	3	Failed to dock with Salyut 1.
Soyuz 11	**6-6-1971**	**24 days**	**6-30-1971**	**3**	**Visited Salyut 1. All 3 died in reentry from asphyxiation.**
Soyuz 12	9-27-1973	2 days	9-29-1973	2	Test of redesigned two-person Soyuz craft.
Soyuz 13	12-18-1973	8 days	12-26-1973	2	Carried Orion 2 Space Observatory.
Soyuz 14	7-3-1974	16 days	7-19-1974	2	Visited Salyut 3.

1968-76 - MANNED LUNAR MISSION PROGRAMS

The Soviet government publicly denied participating in a manned moon landing competition, but secretly pursued two programs in the 1960s: Details of both Soviet programs were kept secret until 1990 when the government allowed them to be published under their policy of *glasnost*.

(1) **The Zond program** – crewed lunar flyby missions (loop around the moon; but not capable if orbiting it) using Soyuz 7K-L1 (Zond) spacecraft launched with the Proton-K rocket.

Soyuz 7K-L1 (Zond)

(2) **The N1/L3 program** - crewed lunar landing missions using the Soyuz 7K-LOK "mother ship" and the LK Lunar Lander, both launched together with the N-1 launch Vehicle.

Soyuz 7K-LOK "mother ship"

The Soviets never accomplished a *manned* lunar flyby, lunar orbit or lunar landing flight with either of these programs.

The Soviet Zond program

The Proton-based Zond program flew 5 flights (Zond-4 through 8) using an unmanned version of the Soyuz 7K-L1 manned spacecraft with biological and later animal test subjects with mixed results. All missions except Zond 4 looped around the moon and the spacecraft for missions 5, 7 and 8 were recovered.

MISSION	DATES	RESULTS
Zond 4	2-7 Mar 1968	Launched away from (opposite) the moon, re-entry guidance failed, capsule destroyed.
Zond 5	14-21 Sep 1968	**First spacecraft to loop around the moon and return live test subjects (tortoises).** Landed in ocean vs. land due to guidance failure; but capsule and test subjects were recovered successfully.
Zond 6	10-17 Nov 1968	Cabin pressure lost killing test animals, parachute deployed early and capsule crashed.
Zond 7	7-14 Aug 1969	First complete success - soft landing with recovery.
Zond 8	21-27 Oct 1970	Soft landing with recovery.

Following the American successes of the first manned lunar orbit flight in December 1968 and the first manned moon landing in July 1969, the Zond program was canceled in 1970.

The Soviet N1/L3 program

The **N1/L3** launch vehicle configuration was developed to compete directly with the United States Apollo-Saturn V to land a man on the Moon, using the same lunar orbit rendezvous (LOR) method as used by Apollo. but However, Apollo used a crew of three astronauts with two of them going to the lunar surface, while the Soviets used a crew of two cosmonauts with just one going to the lunar surface in their smaller LK Lunar Lander vehicle.

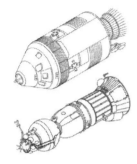

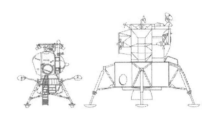

Apollo CSM alongside the
Soyuz 7K-LOK (to scale)

Soviet LK Lunar Lander alongside the
Apollo LM (to scale)

The smaller Lunar Lander was a result of the smaller payload capacity of the N-1 booster when compared to the Apollo Saturn V. Both the Saturn V and the N-1 were three-stage rockets used to place their moon-bound payloads into earth orbit, but the Saturn V third stage could be restarted in earth orbit to place that stage along with the CSM and LM onto a trajectory toward the moon.

The third stage of the N-1, however, was expended upon reaching earth orbit, and the payload's first segment, sitting directly on top of the N-1's third stage, provided the propulsion for the translunar injection (TLI) burn. The payload segments also provided propulsion for MCC, LOI and TEI burns (functionally similar to the Apollo SM), and the last two segments were the orbital and lunar landing vehicles.

The Soviet's approach for their lunar lander (LK) descent to the moon and return to the LOK mother ship in lunar orbit was more complex and risky than for Apollo. The Soviet configuration required extravehicular (EVA) transfers in lunar orbit for cosmonaut access between the mother ship and the lunar lander, whereas for Apollo, the lunar lander (the LM) was *internally* accessible from the Command Module during the entire trip from the earth to the moon, as well as after the return trip rendezvous and docking.

The Soviet configuration required the first EVA transfers for boarding the LK lander prior to descent; the other EVA transfer was on the cosmonaut's return to re-board the mother ship - carrying lunar samples on this trip.

The N1-L3 program was underfunded and rushed, starting development in October 1965, almost four years after the Saturn V. It was also badly derailed by the death of its chief designer Sergei Korolev in 1966. Four unmanned lunar missions were attempted, all unsuccessful due to failures of the N-1 launch vehicle:

February 21, 1969

This rocket was erected on the pad in May of 1968, and everything was finally ready in February of 1969.

By this time Apollo 8 had already orbited the moon but NASA still had a way to go before attempting the landing. There was hope by the Soviets that they could still beat the Americans if this first N-1 launch was trouble-free.

Two N-1 rockets on the launch pad.

However, all 30 engines of the first stage shut down at T+70 seconds and shortly thereafter it was in burning heaps on the ground.

July 4, 1969

The rocket fell back onto the launch pad and collapsed, triggering a series of explosions that engulfed the whole area in flames. It was the largest disaster on a launch pad the Soviet program had experienced, but amazingly no one was killed.

June 27, 1971

This launch started off better than the previous launches but quickly developed roll stabilization problems. This put strong torque forces on the rocket, damaging and ultimately destroying its second stage. Then at T+51 seconds all 30 first stage engines were shut down. The rocket broke apart in the air and crashed.

November 23, 1972

For the first 77 seconds, the rocket behaved as designed and flew further than any of its predecessors. The control system shut down the first stage central cluster of six engines right on time at T+90 seconds, but fourteen seconds later an explosion erupted in the tail of the first stage, and the mission was over.

This last failure was the last hurrah for the N-1 program. After more than a decade of development and eight years of high priority as a lunar landing program, the N-1 mega booster was suspended May, 1974, and in 1976 was officially canceled.

Along with the rest of the Soviet manned lunar programs, the N-1 program was kept secret almost until the collapse of the Soviet Union in December 1991. information about the N-1 was first published in 1989.

PART II

MY JOURNAL - A TRIP TO THE MOON

Reminiscing – looking back - what a wonderful life I've had. It's been a long time since I was a youngster growing up in a small town in the Upper Peninsula of Michigan. Being bussed for 5 years to Manistique for 8th grade through High School, many times only myself and the driver, Bond Tatrow, on the bus. Building, flying and crashing model airplanes – some with the gas engine given me by my Uncle Donnie. Then when in High School building boats from plans in books, first a small 8 ft. racing boat, but no motor so I paddled it around. My dad promised an outboard motor if I built a larger boat that the family could enjoy, so I built a 14 ft. boat and he bought a 7 ½ hp. Mercury. Then I built another 12 ft. fancy boat with windshield and steering wheel up front, motoring into Garden Bay, the open lake to Fayette, Sac Bay and once all the way across Big Bay De Noc to Nahma with Dennis (an uncle only slightly older than me) – trips for fishing but mostly for just having fun. What places, things and changes I have seen since - and then to have had an important part in landing a man on the moon - how lucky. I'll never forget my dad in a serious moment saying "I've had a good life" - me too.

My journey to the moon really started years before I began working in Houston on the moon landing program. It started with my first job in Azusa, California, then continued with the Army in Huntsville and then with General Dynamics/Astronautics in San Diego. Those jobs all involved guidance and control systems for missiles and space systems; each increasing my knowledge and interest in those systems and the discipline of *Systems Engineering*, eventually leading to my acceptance of an offer from TRW to work in Houston, Texas - on the guidance and control systems needed to land a man on the moon.

One of my later assignments at General Dynamics/Astronautics in San Diego was supporting an Apollo proposal team wherein we were the prime contractor and TRW was our Systems Engineering sub-contractor, a somewhat unusual arrangement as usually the prime contractor retained systems engineering responsibility. We lost that competition, but TRW's lead for this proposal, Dr. Erdem Ergin, asked me to join TRW in Houston to work on their Apollo contract. He explained what they were doing there for the Apollo Spacecraft Program Office (ASPO) - trajectory design, analyses and other systems engineering work, and he also described TRW's pioneering Systems Engineering work for the Air Force's ballistic missile program. I was sold, new systems engineering work with NASA that matched my previous work experience on Guidance, Navigation and Control (GNC) systems.

This all sounded great, although abandoning California and moving all the way to Houston with my new family was a somewhat worrisome unknown. San Diego was where I had met and married my wonderful Wilma, who agreed to make this change with our first child Michele, even though we'd be leaving behind her two sisters and other close friends.

Anyway, I accepted Dr. Ergin's offer and joined TRW there – to continue with system engineering type work on America's moon landing program.

1956-57 – CALIFORNIA

Infrared systems, vacuum tubes, and a first
introduction to Guidance & Control systems

It was 1956 and I was ready! A BSEE degree from the University of Michigan, an accepted job offer from Aerojet-General in Azusa, California, and a brand new 1956 Chevrolet convertible to get me there. It was all good, I would be working on infra-red tracking and missile

1956 Chev – San Diego with Jack & Celia Fox

guidance systems with a great salary of $490 per month. I had borrowed the $300 down payment from my mother and with Uncle Donnie's connections bought my first car in nearby Detroit at a good price of $2,500.

I was off to my first real job and the fun, sun and excitement of Southern California.

At Aerojet, I worked on an infra-red (IR) tracking system for a surveillance aircraft and an IR seeker system for the Sparrow air-to-air missile, doing analysis of theoretical search performance and tests of detection and tracking performance for the lead sulfide (detector) system. I still have a "Radiation Calculator" slide rule from General Electric which I used for this work (a valuable museum piece?). Hands-on work with electronics included breadboard construction and testing of an airborne power supply - vacuum tubes were used then. And I wondered about the reliability of the Sparrow missile after looking inside the 8" diameter missile and seeing the canister shell almost completely covered with subminiature vacuum tubes, each tube with a fragile heater, cathode, grid and plate, but at this time only one transistor had recently been approved for military systems.

1957 – Into the Army, but not exactly as planned

The Southern California sun and fun lasted only one year. In the summer of 1957 I was called-up to be drafted, with orders to report to Pasadena for induction into the US Army for 2 years of service at Private E-2 rank. I went to Pasadena as directed and took and passed the physical and written exams. The recruiting sergeant expressed surprise when I correctly answered all exam questions – not much was apparently expected for draftees at that time; just a reasonably capable warm body.

But I was not eager to leave Southern California to enter military life, so I left Aerojet and drove from Azusa to Michigan to petition my local draft board in Escanaba for a deferment. It didn't happen, so I notified the Pasadena draft office that it was inconvenient for me to report for induction there as I was in Michigan; they granted me a delay.

1956 Chev – at parent's home in Northern Michigan. Next – Phoenix, AZ

Realizing that my $85 monthly car payment would exceed the $78/month Private E-2's pay, I drove to sunny Phoenix in Arizona – a better place than Northern Michigan to sell a convertible in the fall, and besides, Phoenix was on the way to Pasadena where I was scheduled to report for induction. In Phoenix I stayed with a favorite, Aunt Lorna - I had lived with her in Tucson for my first year of college at the University of Arizona. While there, though, I again contacted the Pasadena Draft board stating that I would now report for induction in Phoenix. Again they had little choice as I was already there, so they granted me another delay.

Anyway, the car sold in Phoenix, and when I went to the Phoenix induction center I was told that because of my college degree instead of being drafted I could enlist into the *Regular* US Army for a 3-year period and would become a 2nd Lieutenant after basic training. Three years as an Officer sounded much better than 2 years as a private, so on September 9, 1957, I enlisted for a three-year term in the Regular US Army and went in as a private E-2 with my officer's commission to be awarded after completion of basic training.

This didn't work out exactly as planned; I ended up in the Army as a Private for 3 years because the officer's commission didn't come through until about a year later when I was in Huntsville, and when it did, it required I spend a total of 3 years in the Army as an officer. After having been in Huntsville for almost a year, that would have been 4 years total, so I declined the officer's commission and continued on as a Private. Besides, after one year in Huntsville I had made several good army friends – all draftees, except Jack O'Loughlin, and all of private rank as well. I was also enjoying my work with the civilian (German) engineers – almost just like a regular job. Another advantage I later realized was that being a Regular Army enlistee rather that a draftee I was favored for promotion, first to the rank of corporal, and later at the end to Specialist 5 - equivalent to Sergeant. More money - $180 per month - and even my own private room in the barracks near the end!

1958 - IN THE ARMY – at a Nike Site

How did I end up in Huntsville, Alabama, of all places? It was Jack O'Loughlin's fault, but he ended up there too, and it was all good after that – lucky again.

I first met Jack on a train taking us to our first assigned duty location after 8 weeks of basic training at Fort Ord. Because we were in uniform we had started talking, found out we were both engineers and were going to the same place, a Nike Ajax Battery at Fairchild Air Force Base near Spokane, Washington. Apparently the Army thought this assignment was the best fit for us engineers - shooting down airplanes with missiles.

Anyway, while at the Nike base I read in an army publication about their Scientific and Professional (S&P) program for individuals with a technical degree. I mentioned this to Jack and we both applied to base administration to change our MOS (Military Occupational Specialty) as needed to be transferred into the S&P program. Long story short, it was all approved, we were transferred into the S&P program. However, with no openings for this MOS at the Nike site, we selected Redstone Arsenal out of a few candidates because Jack's friend Lieutenant Joe Ventura was stationed there. The warmer climate would be a welcome change from the wintry cold and snow of the Spokane area.

NIKE SITE RECOLLECTIONS

Banks of vacuum tubes (maybe hundreds of them) glowing inside the Nike computer cabinets. Very unreliable, we replaced tubes often. Remember, no solid state electronics then.

Me – on guard

Training exercises where we pretended to shoot down passing US military aircraft flying by within the range of our missile. An alarm sounds, often in the middle of the night, we get up, get dressed and run out into the snow and cold to quickly align and calibrate the Missile and Target Tracking Radars so accurate steering commands can be calculated and sent to the missile for a successful intercept. Of course we never launched the missiles, but for these exercises the missile steering commands went to a missile simulation computer (more vacuum tubes) so our performance could be evaluated.

White Sands Trip On one occasion our battalion was loaded into two planes (CIA's Air America, we thought) that took us to White Sands, where we actually launched the Ajax missiles and shot down target drones. An interesting trip, especially the poker game in the bouncing back of the plane on the way down. Many were sick and vomiting from the severe turbulence, but I was OK and did quite well at the poker game. For that I thank my Greene relatives for their many instructive poker sessions, and fondly my early start with penny-ante poker on the dining room table after Sunday dinners that included us kids as well as the adults.

Two airplanes took us there – my plane was OK but encountered severe turbulence. The other plane had engine trouble, had to be replaced and they arrived the next day. The officers were apparently bothered also by these flights, so they arranged for all of us to return to Spokane on commercial flights instead of the military charter.

1958-60 - HUNTSVILLE – with the ABMA

working with the Germans, another infrared system, and an improbable relationship of trust

At Redstone Arsenal Jack and I were assigned into Detachment B of the Army Ballistic Missile Agency under General John B. Medaris – and we were in the S&P program. The Space Race was well underway when we arrived; the USSR had launched the first earth-orbiting satellite and the ABMA had already developed the Redstone and Jupiter missiles supporting Wernher von Braun's early efforts and launched the first US satellite into earth orbit.

ABMA
Shoulder Patch

The biggest issue existing at that time was how best to protect a ballistic missile warhead or a manned spacecraft capsule to survive the high heat experienced during their re-entry into the earth's atmosphere. There were only two heat shield approaches, a beryllium heat sink or a heat shield made of an ablative material - fiberglass bonded with a resin. In those early days of uncertainty, the Mercury capsule design was modified to accommodate either type; the ablative approach was finally used due to its lesser weight.

I was assigned to Dr. Haeussermann's Guidance and Control Laboratory, working for Jim Reinboldt, Chief, Applied Research Test & Evaluation.

Similar to my work at Aerojet-General, my major project in the ABMA also involved an infrared system - the development of an infrared horizon scanner. Two of these were used with each Mercury spacecraft to provide roll and pitch orientation signals so the capsule's attitude control system could keep the astronaut and his capsule in a stable orientation similar to an airplane flying high above the earth.

These horizon scanners are shown here - in the nose of this Mercury spacecraft cutaway.

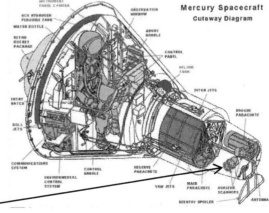

Mercury Spacecraft
Cutaway Diagram

The following 2004 e-mail, sent to my granddaughter for her 8th grade rocket project, summarizes my technical work in Huntsville:

Dear Anna,

Here is a picture of me, PFC Richard Boudreau, on assignment in Washington DC standing in front of an actual Jupiter C Launch Vehicle. Myself, one other PFC and the Jupiter C launch vehicle were an exhibit at an AUSA (Association of the United States Army) Symposium in late 1958.

My job at the symposium was to explain to the attendees the Army's role and accomplishments in America's space program.

When the Soviet Union launched their Sputnik satellite into earth orbit in October 1957, the "Space Race" was on. Just 118 days later, on 31 January 1958, the United States used the Army's Jupiter C to launch into orbit America's first earth-orbiting satellite, Explorer 1.

Under the direction of Dr. Wernher Von Braun, the US Army Ballistic Missile Agency (ABMA), located at Redstone Arsenal in Huntsville, Alabama, developed the Jupiter C launch vehicle. Von Braun's team also developed other space systems and the much larger launch vehicles needed for manned earth orbital flight and for landing a man on the moon.

While in the Army, I was categorized as one of the Army's Scientific & Professional personnel. We were all enlisted men with technical, scientific and engineering degrees. We were in the ABMA, but assigned to support Dr. Von Braun in his space systems and launch vehicle development efforts. It was almost like a regular civilian job. Every weekday we went to the engineering laboratories to do engineering and scientific work. I was the Project Engineer for development of a horizon scanner that was used on America's first manned space flights. I designed and developed test fixtures, performed laboratory development and environmental testing (heat and vibration tests) of these scanners, and traveled to the manufacturer's site (Barnes Engineering in Stamford, Connecticut) for technical coordination and to resolve issues. I also went to Cape Canaveral, Florida for flight testing of these units. All this was a very interesting and educational experience for me.

Hope this helps with your project. It was fun for me to locate this old photo and write this brief summary of my Army experience.

Love, Papa

During my Huntsville days I made two other lifelong friends besides Jack O'Loughlin - Carlos "Charlie" Navarette and George Anas. To this day the four of us get together with our wives for fun times, which include of course reliving old Huntsville events.

It wasn't all technical work in the Huntsville laboratories – there was "army" work as well. We were told that even though we were in the S&P program, our primary MOS was still infantry. So every Wednesday morning we had physical training/exercises, training movies and marching around – and occasionally marched in a parade as well. And of course there were periodic barracks cleaning "parties" to prepare for inspections, and regular "KP" (Kitchen Police) duties like washing dishes, peeling potatoes, and so forth – we had to do whatever we were told to do (occasionally we paid others to do our KP duty).

There were fun times as well that may (or may not) have helped in our growth. Available were NCO Club nickel-dime nights, "9 for 5" vacations (9 days off for 5-days charged against our annual leave) and bottles of cheap scotch and bourbon for trips to Mobile Alabama, New Orleans (Mardi Gras), Panama City, etc. We had very little money, so we slept in our cars, on the beach, or wherever during these trips (some of these stories will not be told – they'll remain in the "cone of silence").

We also had use of the on-base auto shop with its many tools. Shortly after arriving in Huntsville I used that shop with its engine lift to install new crankshaft bearing inserts and rebuilt the carburetor for a $200 1951 Ford I had purchased in

Jack & I – we slept here in New Orleans

Detroit – went thru Uncle Donnie again as I did before for the '56 Chevy Convertible. This one though was kind of a "junker" with much rust hidden by a new paint job. Shortly after arriving in Huntsville with the car, the left front fender sprung loose from the frame, and then I found under the rear floor mat a hole in the floorboard covered by an old license plate – the hole large enough to drop a beer can through - but the price had been right.

Jerry Lombardi, a young engineer like myself in the S&P program, introduced me to Dr. John Cashin, a dentist in Huntsville. A nice relationship, he would take his boat to Guntersville Lake and let a few of us ABMA guys use it for water skiing. Jerry and I replaced a broken camshaft in his Porsche and in return he loaned it to us for a trip to Acapulco and back. About 2,000 miles each way, this was very generous of him and a great experience for the two of us. I had gotten to know Jerry well from visiting his parents during my Barnes Engineering trips to his home town of Stamford, Connecticut.

Jerry - on Dr. Cashin's boat and beside his Porsche in Mexico

Jerry is the one who first met and got to know Dr. Cashin - I assisted him in replacing the camshaft; met Dr. Cashin only once or twice casually while working on his car. I do not know how he and Jerry connected-up back then, in the late 1950's, a black dentist and a white soldier assigned to Redstone arsenal, and how might he have gotten to know Jerry well enough that he would trust him with his Porsche sports car for such a long trip?

Richard – somewhere in Mexico on our Acapulco trip

My online search did not find Jerry, but did find two Dr. John L. Cashins in Huntsville who were both dentists; Dr. Cashin Sr. and his son Dr. Cashin Jr., born April 1928, the Porsche owner. After the junior Dr. Cashin received his D.D.S. degree in 1952 he was shortly after drafted into the Army and served his 2 years in France, returning afterwards to practice dentistry with his father in Huntsville.

There is much online information on Dr. Cashin Jr. Sometime after we had met him, he became very active in the civil rights movement and founded the National Democratic Party of Alabama in 1968 to oppose the extremely anti-integrationist Democratic Party embodied in the region by Governor George Wallace[6], and ran (unsuccessfully) in 1970 for governor of Alabama against Governor Wallace. The party he founded, however, did succeed in changing the face of local offices throughout the state. In 1968, his party's candidates won 17 local offices in the central and western part of the state, and in July 1969, six black candidates won seats in Greene County's government, the first time since 1819 that that county was not governed exclusively by whites. In 2009, *The Huntsville Times* in Alabama called Dr. Cashin "one of the most ferocious civil rights lions in Alabama back in the day."

---

[6] How far we have come in 50 years since the openly racist George Wallace won 13.5% of the popular vote in his run as a third party candidate for US president!

I did connect up via e-mail with his daughter Sheryll Cashin – a Professor of Law at the Georgetown University Law Center - to hopefully obtain insight into how, in 1958-1960, Dr. Cashin might have connected-up with Jerry. Long story short - an excerpt from her return e-mail:

> The Dr. Cashin you met was, indeed, my father. He was something else. An agitator, brilliant polymath with deep interests in many things including rocket science and a passion for civil rights. If you read my family memoir, The Agitator's Daughter, you will have the full context and a sense of how it could be that a black dentist would have such a relationship of trust with your friend Jerry that he would loan his car to him and take him skiing on his boat.

> I have no recollection of the Porsche. I was born in 1961. But I do remember the boat and lots of other toys my father had, including an antique Rolls Royce. I have never heard of Jerry. My dad went to the Unitarian church in HSV and met a lot of scientists and engineers there. We often had such folks over to our house and some became great friends. "Everything else was so boring" he once said to me. My dad liked exploring things and having fun especially before he got very serious about civil rights.

I continued my efforts to find Jerry, and surprise, in December 2018 he called me in response to my months-ago Facebook message. He was the Lombardi I was looking for, and healthy and cognizant! You never know, it's been 60 years with no contact since we were in the army together.

Anyway, he told me the story of how he got to know Dr. Cashin to the point where he would trust him with his Porsche for our Mexico trip.

It all started with Jerry's bad tooth and his lost model airplane.

Jerry had a cavity, and being in the army went to the on-base dentist to have it filled. After application of the Novocain, the dentist started work on another patient, ostensibly waiting for Jerry's anesthetic to take effect. After too long a time, however, the dentist came back to him and started drilling on Jerry's tooth, but the Novocain had worn off. A very unhappy Jerry got up and left – mid-procedure.

But the tooth still needed drilling and filling, so going to the yellow pages, he found a Dr. Cashin and had him do the work - all good now with the tooth.

Jerry also was a Radio Controlled (RC) model airplane enthusiast and a few months after having fixed his tooth, one of his model planes flew out of radio range and was lost. But a note in the plane with Jerry's contact information led to it being found by a Dr. Harold Drake, who also built model airplanes and turned out to be a close friend of Dr. Cashin.

Long story short, this led a mutual friendship of almost two years, with Jerry building RC model airplanes on the pool table in Dr. Drake's basement and replacing the camshaft in Dr. Cashin's Porsche - with my help.

The Acapulco trip was like many others taken while in the army, taking turns driving to shorten the trip and mostly sleeping in the car to save our money for gas and food. For emergencies only I did have a Mobil Oil credit card - used once to buy a "tire" where I received $20 instead from a cooperating Mobil station. The card was arranged for me by my dad and his friend who owned a Mobil gas station.

Not to sure of our camshaft repair job, Jerry red-lined the engine before entering Mexico - better to fix it in the U.S. if needed.

More drama in Alabama

After reading a *Time* Magazine article[7] about how to add drama to your life story, I decided to include three of what I will call my "near death" experiences. This is somewhat an exaggeration, they were really just alarming situations I found myself in, but labeling them "near-death" hopefully adds a little more drama to this narrative.

Anyway, my first "near-death" experience took place in Huntsville, when four of us army guys were quietly enjoying our beer in a sawdust-floor bar down by the railroad station. We were in uniform, therefore obviously not local, which may have made the bartender somewhat nervous as other soldiers from Redstone Arsenal were typically more rowdy and aggressive than us when out drinking (the MPs come to mind). Anyway we were ready to go on to the next place, so we all got up and started leaving but I was somewhat behind the others in the drinking, so I was leaving with my glass half-full of beer. As we were walking out, the bartender pulled a 45 caliber pistol from

---

[7] Oct 12, 2015 *Time* magazine article; "The art of the boring memoir; or how I learned to find drama in my life story" by Joel Stein, page 63.

under the bar, pointed it at my head and said "that glass belongs here." Needless to say I set the glass of beer back onto the table; the first and only time I have ever had a gun pointed at me with seemingly threatening intent – an alarming experience for my first time in the "deep south." At that time we believed that the locals perceived us as lower than snakes - they didn't like us when we were out and about – especially with "their" local girls.

My active duty in the Army ended June 14, 1960. However, my civilian boss Earl J. Reinbolt said he wanted me to stay on in Huntsville, likely at a GS-12 level, working either in his section or elsewhere in the G&C lab. He worked for Dr. Jasper in Dr. Hausermann's Guidance and Control lab. But I was anxious to leave the army life and return to California and find employment there, so at my request he wrote me a very nice letter of recommendation – of interest here in that it is his perception of the job I was doing. The GS-12 salary was tempting, but in my mind Huntsville life could not compare with California.

Mr. Earl J. Reinbolt went by "Jim" - a very nice man and a pleasure to work for.[8] In connection with our work, he sent me on three trips that I remember – two to Barnes Engineering in Stamford, Connecticut, the developer and manufacturer of the Mercury horizon scanners, and one to Kennedy Space Center (KSC) for flight testing one of these scanners.

On the trips to Barnes Engineering, I would fly to New York City, stay overnight at the YMCA, and next day take the train to Stamford. It turned out that Jerry Lombardi's parents also lived in Stamford, so on one of my trips I stopped by their home and Jerry's father gave me 2-3 bottles of red wine that he had made and stored in his bomb-shelter basement; to take back to Huntsville for his son and friends. On the way back however, one of the bottles had leaked - I ended up with a red-stained suitcase, but 90+% of the wine survived and was later consumed with appreciation. On another trip when in New York, I obtained theater tickets and thoroughly enjoyed "Hernando's Hideaway" "Steam Heat" and other numbers while attending the stage musical *Pajama Game*.

---

[8] Thinking back on these long-ago events, a small regret today is somehow not getting some of Jim's personal story – where he grew up, family or other ties in Huntsville, etc. - but I was young.

My trip to KSC to flight test the horizon scanner was likely sometime in 1959 or 1960, about 2-3 years before the first Mercury flight in February 1962. I drove round trip from Huntsville to KSC in my 1954 Buick (ABMA provided mileage reimbursement), taking with me a test fixture used to ensure proper operation of the scanner after its installation in the Jupiter missile. I had designed and had built this fixture in the Huntsville Labs - did the drawings, optical ray tracing, testing, etc. This fixture had a large circular concave mirror to reflect IR radiation into the scanner aperture – the IR radiation was generated by a heated disk segment at the mirror's focal plane to simulate the earth's infra-red radiation temperature.

At KSC, after going up on the launch gantry elevator, I accessed the instrument compartment opening near the installed horizon scanner, installed the test fixture on the outside covering the aperture of the horizon scanner, and then connected an electrical cable to the scanner on the inside. While inside on top of the fuel tank (it was empty), it sprung inward somewhat due to my weight; but I was relieved that when I backed out it sprung back into place. My first thought was I had damaged the tank, or at least overstressed the tank-top metal. I told a responsible person who said don't worry about it, it should be no problem.

To my relief the launch went without any problems, and post flight analysis of the telemetered data showed the horizon scanner worked as expected and that the earth radiation signal levels were also as expected. This was the first time the earth's radiation level in the range of the lead-sulfide detector's wavelength region had been measured against the dark space background!

JIM's LETTER OF RECOMMENDATION

N A T I O N A L
A E R O N A U T I C S
A N D   S P A C E
A D M I N I S T R A T I O N

IN REPLY REFER TO

GEORGE C. MARSHALL
SPACE FLIGHT CENTER
H U N T S V I L L E, A L A B A M A
TELEPHONE: JEFFERSON 6-4411

August 13, 1960

TO WHOM IT MAY CONCERN:

Mr. Richard Boudreau was employed in my Section from April, 1958 to June, 1960. During this period he proved to be an extremely capable person both in the technical sense and in assuming responsibilities willingly. He also displayed the ability to deal with other people with a minimum of friction. I recommend him to you without any reservations.

During his period of employment here he was engaged primarily in the design and testing of optical and infrared sensing devices and their associated circuitry. Part of his assignment included the study of how these devices could be integrated into overall missile and satellite guidance and control functions.

Even though Mr. Boudreau was in the U. S. Army at the time, he functioned as a full time employee and assumed the same technical responsibilities demanded of civilian employees in corresponding positions. Personally I would have been happy to have retained the services of Mr. Boudreau; however, he displayed great interest in locating in your area.

EARL J. REINBOLT
Chief, Test and Evaluation Section
Applied Research Branch
Guidance and Control Laboratory

TRANSCRIPT OF JIM'S HANDWRITTEN NOTE

August 16, 1960

Dear Dick,

I have enclosed the original and two copies of the letter you asked for – hope that it is helpful to you. In any case you can give my name as reference – I will be glad to answer inquires from your prospective employer.

Sorry we missed each other when you were leaving, but I'll now wish you the best of luck in your efforts in California. I'm sure you'll meet with success. If not, let me know and send me a Gov't Form 57. I'm sure I could find a place for you either in my section or elsewhere in G & C lab (at GS -12 level probably).

I saw Lusch just recently (he is from Advanced Technology Lab). If I had known then you were looking perhaps I could have helped some.

It is too bad you were not able to finance your school effort. It certainly would have been easier to do it all at once than to do it with night courses, but eating is a necessary requirement also. I'm sure you will continue your studies at your new job. Nowadays if you want to keep up with your field, it seems that continual study is a requirement.

Let me hear from you. I would be interested in knowing where you finally settle. Like you say, I do take trips and if I come to your new neighborhood I'll look you up.

Sincerely,

Jim

Mandatory National Service

My time in the army was overall a very positive experience. I believe every young person out of High School or College should be required to perform one to two years of mandatory service for our country; good for our country and a good time for that person to grow up and obtain a better understanding of the real world. The following article from the September 21, 2016 *Honolulu Star-Advertiser* expresses these sentiments well:

Mandatory national service would strengthen America

By Victor Craft and Peter Adler

In the late 1960s, Victor Craft was an Air Force enlisted staff sergeant in Vietnam fixing aircraft that had been shot up, while Peter Adler was in India with the Peace Corps raising chickens, building schools and killing rats.

America in general and Hawaii in particular need a new national service requirement in which every able-bodied young man and woman is required to commit one or two years of time to a cause larger than themselves. We have had this before in the face of external threats. Today the threats are internal. We need citizens to have more skin in the game.

Why? Citizenship may appear to expand with every new law but ironically, it is shrinking. Everyone is focused on the cult of "Where's mine!" Interconnectivity has many advantages but it also seems to increase political pessimism. Our sense of community is reduced to laws to be obeyed, votes to be harvested, and taxes to be paid.

Is there a cure? A new social contract with tangible expectations and opportunities could be a start. Having ourselves volunteered during a different era of American politics, we can see how those experiences shaped new disciplines and habits. Even though we come from different traditions, one military and the other civilian, it created a shared civic experience.

Here is a better example.

In 1939, a law passed by Congress and Franklin Roosevelt changed the temperament of America. It created what would soon become the Works Projects Administration, an ambitious service program that put millions of men and women to work to carry out projects that included the construction of parks, roads, bridges and buildings.

Many of those constructions endure today, but there were other, maybe even more important legacies: life-changing odysseys for young people; tangible work that got done with pride; and the direct involvement in what politicians now call "The American Dream."

Congress also passed the Selective Training and Service Act, the first peacetime conscription in U.S. history. Selective Service required men between 21 and 35 to register with local draft boards and be ready for call-up. Two decades later the original architects of Kennedy's Peace Corps program envisioned the same. A Peace Corps draft.

Today, we need an experience of direct citizenship to help counter the loss of faith in our institutions. Citizenship without meaningful participation makes us lazy. It becomes a constant assertion of "rights" devoid of "responsibilities."

There is much that needs to be done. Our state and national infrastructures are in dire need of maintenance. Schools need additional teachers. Poorer communities could use assistance from young lawyers, MBAs and social workers. Parks and trails need to be fixed and start-up minority businesses and social service agencies need entrepreneurial help.

To stay healthy, we need fresh imagination and political willpower and a focus on "doing" rather than talking. There are plenty of options available: any branch of the military, Teacher Corps, Vista, AmeriCorps, Peace Corps, and YouthBuild. More can be invented.

Hawaii with its general civic apathy could especially benefit from programs that motivate 18- to 25-year-olds to do something for the community. We can also create "carrots" — college debt relief; post-volunteer career or college scholarships; hiring preferences; re-adjustment allowances after volunteered time.

The long-term payoff could be profound. "The best way to find yourself," said Gandhi, "is to lose yourself in the service of others." Much of this comes down to inspired and brave leaders who will insist on a service requirement and deliver up projects and programs that people can touch.

[end of *Star-Advertiser* article]

1960–65 SAN DIEGO

More GNC work – Wilma enters the picture

After interviewing several California companies, in late 1960 I joined General Dynamics Convair Astronautics in San Diego. I first worked in Krafft Ehricke's Advanced Systems Department doing conceptual design work on the Air Force's Manned Orbiting Laboratory (MOL) and early moon landing concepts (one for example used multiple launches and earth-orbit rendezvous to assemble the many Centaur stages needed for adequate propulsive capability – a very long and thin vehicle. These studies were accomplished in the very early days of computers; we had a GE 600 series machine. In support of the MOL project for example I wrote a few simple programs for this machine to calculate optical visibility parameters from an earth orbiting vehicle. This coding work used mnemonics (abbreviation codes) to represent the low-level machine instructions needed - all very primitive compared to my later FORTRAN programming.

After that I worked on the Atlas Centaur guidance system, doing lab testing, error modeling and analysis of the Atlas/Centaur inertial guidance system performance.

From my 5 years at Convair, I made several good and long-lasting (engineer) friends: Ray (Glenna) Thompson, Jim (Helga) Moore, Matt (Solfrid, then Raquel) Nilson, Bob (Beverly) Cramb, and Harry (Mary Carol) Eastman. And at a movie theater (with Wilma then) ran into and reconnected with Charlie Navarette from my army days in Huntsville; he and his wife-to-be Bobbie were also working for GD Astronautics in San Diego at that time. Later Ray Thompson took a job with NASA MSC and we hosted him and his family at our home in Houston for a few weeks until they moved into their home there, and more recently we visited with them here in Hawaii when they were on assignment working at the Polynesian Cultural Center for the LDS Church. We still often attend Matt's annual birthday party in San Diego, which keeps us in contact with our "Convair group" friends.[9]

---

[9] In June 1961 the Convair-Astronautics division of General Dynamics name was changed to General Dynamics Astronautics. Many of my friends and I still use the name "Convair" to describe where we worked at that time.

It was in San Diego that I luckily met and married Wilma, my travelling companion and the stimulating spice in my life. As anyone who knows her will attest, she is the hard-working and enthusiastic social director, organizer and manager of our family life, keeping us all in frequent contact - as well as continuing to locate new friends and relatives with her Portuguese genealogy work. She exudes excitement and gives much happiness to all who know her.

My first sight of Wilma was when she was in the swimming pool of her San Diego apartment complex. I had driven my Convair engineer friend Matt Nilson home to his apartment, which just happened to be in the same building where Wilma lived, in an upstairs unit with her

1963 – Our Wedding

sister Duanne. Anyhow, I invited her and her friends to go water skiing, she accepted and we did that. To make a long story short, while skiing on one ski, I was going to show-off my expertise by swooping grandly up to the beach and calmly step off the ski near where Wilma was sitting. Stepping off the ski and walking onto the beach was easy, no problem as I had done this before, but this time the ski caught solidly in the sand and I fell forward flat onto my face. She was not impressed and I had cracked a rib.

Waterskiing at Mission Bay, San Diego

Later on she told me that while working at her job in an insurance office she saw the claim for my rib injury come through and, along with other somewhat personal information, decided I was an acceptable dating candidate.

After that things were great in all respects; much fun with our friends at parties, boating, trips to Palm Springs, a trip to Las Vegas with Aunt Lorna, several trips to Tijuana, etc. Our first kiss was on the dance floor of the Capri Club in Tijuana. So all was going very good until sometime later when we were talking in my Corvette in front of her apartment complex she told me she was thinking of returning to Hawaii! I got off the dime and proposed, she accepted, but said she wouldn't marry me until she was of legal age, that way her strict father would not be able to stop the wedding. We were married February 23, 1963, 15 days after her 21st birthday.

Nine months later our first child Michele was born - in December 1963 - a major milestone of course. How lucky today we all are to have her and her extended family members in our lives.

Honeymoon in Hawaii　　　　**in our first apartment**

1964 - Michele's favorite place

Wilma & Michele in front of our　　**she crawled up there herself**
first house – in San Diego　　　　**- and never fell**

A Plane Trip East

Back to trying to add some drama to this story. Both my second and third "near-death" experiences occurred during a plane trip from San Diego to Michigan and back - before we were married.

Wilma was the caretaker of my Corvette (she had a good parking place near her condo) while four of us Convair engineers took a single-engine plane leaving from San Diego flying east to New York City, on the way dropping each of us off at various locations to visit family and friends, then picking us up on the way back. Matt Nilson was our pilot and we four shared in the rental cost of the plane - a 4-seat single-engine Mooney M-20. Its wings and tail empennage was a wood framework covered with cloth that reminded me of the paper-covered balsa-wood model planes I had built as a youth – but likely OK for real planes. The plan as carried out was that first, Marshall a friend of Matt's was dropped off in Minneapolis, then I was dropped off in Escanaba - in Michigan's upper peninsula, and then Matt and Harry flew on to east-coast points. We were then picked up in reverse order for the return to San Diego, with Harry being dropped in Denver for an extended stay.

A good plan and not to worry, Matt our capable pilot was certified for Instrument-Flight Rules (IFR), which means the FAA allows him to fly "blind" using only the plane's instruments - into clouds or at night with no outside visual cues. And not too long a trip and not too many stops - the Mooney cruised at about 160

Matt - our intrepid pilot

mph and needed refueling stops only every 500 miles or so.

After dropping Marshall off in Minneapolis, we left for Escanaba with a refueling stop in Marinette, Wisconsin, on the way. At the Marinette airport we had to fly through a cloud bank at the downwind end of the runway for our landing approach, but no problem, Matt was IFR rated and certified to fly through clouds. Upon exiting the cloud bank on our first approach we were not lined-up with the runway. So Matt went around again and on this second try we came out of the clouds properly lined-up and landed just fine.

Locals in the terminal building watching our landing attempts told us they were quite concerned we were going to hit a tall paper-mill smoke stack hidden in the clouds near the end of the runway.   After a while the clouds lifted. Matt, Harry and I will likely never forget seeing that high smokestack we luckily had missed – certain death for sure if we had.

Leaving Marinette and heading north, after several unsuccessful attempts to find smooth weather higher up we flew low, following railroad tracks to arrive safely in Escanaba for a short visit with my parents – and to tell them myself of our wedding plans (and that Wilma doesn't wear a grass skirt).   They were not that enthused, but had never met Wilma either.

In Northern Michigan

My third and last "near-death" experience was on the return leg of this trip. It was a beautiful and clear night over the desert landscape, a few far-distant clouds with lightning in them. We were flying west toward Albuquerque at about 10,000 ft. altitude, the maximum allowed without supplemental oxygen. About 85 miles from our planned stop in Albuquerque, I noticed that the oil pressure gage was reading zero – how long at that value I didn't know.   I thought – if the engine freezes-up and quits now, we could glide only about 12 miles before coming down to earth (I kept track of these things - multiplying 10,000 ft. by 6 - the plane had a 6 to 1 glide slope - equals 60,000 ft. divided by 5,000 ft. per mile roughly equals 12 miles).   But we were about 85 miles (30 minutes) out from Albuquerque, so if the engine should quit where would we land?   We were over a desert region, and making our situation worse Albuquerque is surrounded by high mountains.

Looking down though, I saw below us the lights of Las Vegas, New Mexico (who knew there was another Las Vegas?) and it had an airport! So I suggested to Matt that we land there ASAP, hopefully before the engine quit. He overrode that; the engine was still running and he thought that overnight accommodations would be much better in Albuquerque, so he figuratively wrapped a scarf around his neck, tightened his aviator goggles and we flew on. Luck again was with us; the engine continued running for the next 30 minutes or so and we landed safely in Albuquerque.

After landing, we saw oil all over the belly of the plane, and found that only about 1 quart was left - we had lost 7 of the 8-quarts capacity. Next day the Albuquerque repair mechanic found that the repairs made on our flight director instrument by a Minneapolis mechanic had resulted in the oil leak.

That darkness on the bottom front of the plane is oil

1965–74 HOUSTON – Apollo Mission Techniques

with Wilma, Michele & Kathleen. Roland arrives later

We arrived in Houston in June of 1965. My first memory of Houston was viewing the steaming fields across the Gulf Freeway from the front window of our Holiday Inn - Houston was definitely hot and humid justifying our San Diego decision to install an air conditioner in the 1964

Off to Houston in our '64 Ford

Ford which had taken us there – to the Manned Spacecraft Center of the world!

Four years after President Kennedy announced our nation's goal of flying to the moon, I had resigned from General Dynamics Astronautics in San Diego to work on the moon landing program - for TRW Houston Operations under TRW's engineering support contract with NASA's Manned Spacecraft Center.

With Wilma and Michele – she was 1 ½ years old then - the three of us departed San Diego and drove to my home town of Garden in Upper Michigan to pick up my sister Kathleen. She would be living with us, then later near us, for the next 9 years attending college, working as a school teacher and, as it turned out, meeting and marrying her husband Tommy Cox.

Just two years after our marriage in 1963 exciting times were ahead for us all – we would be a part of our nation's quest to fly men to the moon and back! We had arrived near the beginning of the Gemini Program - earth orbit flights of a two-man capsule to develop the capabilities and systems needed for long-duration space flights, an astronaut's ability to work outside in the space environment and techniques for rendezvous of two orbiting spacecraft – all essential needs for the Apollo program.

We rented an apartment for a few months in League City but then purchased our first home, brand new at 1910 Fairwind Road, where we would live for the next 9 years. Exciting, but a nervous time too – our biggest debt ever (at that time).

1910 Fairwind Road – our first home

Apollo Mission Techniques

My major role in the Apollo moon-landing program was manager of a 3-year TRW effort supporting NASA to develop and document "Apollo Mission Techniques." These techniques were described in documents authored by my TRW team that described how all the Guidance, Navigation and Control (GNC) systems, both on the ground and onboard the spacecraft, were to be used, together, in controlling the Apollo spacecraft trajectories to the moon and back. Given the operational complexity of the Apollo mission and the number of GNC systems involved, this was an imposing task.

The Apollo Mission Techniques documents defined a *baseline plan* that controlled the efforts of many hundreds of engineers doing engineering analyses and development of the detailed flight plans, procedures, and mission rules for the Apollo manned missions. They were formally recognized by NASA management as having precedence over Apollo Mission Rules, Flight Crew Procedures and Flight Plans.

Typically well over 300 copies of each document were distributed to the many involved government and contractor personnel – Astronauts, MCC-H ground crew personnel, spacecraft and software developers, and MSC in-house engineering and mission planning organizations.

TRW's early work

In early 1966, Marvin Fox, my supervisor and manager of TRW's Houston Operation's Guidance and Control Department, identified the need for a "data priority" effort based upon TRW's Gemini post flight analysis work. This was briefed to Dr. Joseph Shea, ASPO Manager, who then initiated a two-man TRW task reporting directly to him. Dr. Rosenbloom (TRW Houston Operations manager), Marvin, myself and later Bob Kidd were the initial performers on this task, and I was assigned as task manager (Task ASPO 46).

Working with other TRW and MSC engineers, my TRW team went on to develop and publish, for MSC, the first "Apollo Mission Techniques" document shown here – covering data priority issues for the two most critical Apollo mission phases, Lunar Descent and Ascent/Rendezvous.

Logic Flow Charts were the key presentation method - long fold-out charts that show astronaut and ground crew functions, data flow between them, and decision logic for key real-time GN&C decisions. These charts were the catalyst for many meetings between TRW engineers and appropriate members of the later Tindall mission phase teams.

The TRW document dated September 12, 1967 titled "Apollo Mission Techniques" has an Introduction that contained the ground rules for developing the mission techniques and constructing the logic flow charts, and completed sections for the Ascent and Rendezvous mission phases. Placeholder dividers were used for the Earth Launch, Trajectory Maneuvers, Descent and Reentry mission phases.

We published the Descent chapter of this document one month later, on 16 October 1967.[10]

The Apollo fire - During a routine prelaunch test on January 27, 1967, The Apollo program changed forever when a flash fire swept through the Apollo 1 command module, killing all three astronauts. Instead of launching America's first manned Apollo flight, the program was placed on hold while major modifications in the command module design, materials, and procedures were implemented, including a Block II Command Module.

---

[10] Originals of these two documents and a Gemini Rendezvous Study document were donated to UHCL Archives in October 2015.

Bill Tindall

After the fire, Howard W. "Bill" Tindall, Jr. was appointed Chief, Apollo Data Priority Coordination - in August 1967. Shortly after his appointment, he requested that we publish these two documents as a starting point for his subsequently planned effort, which we did.

With the first of these two documents attached, he then issued, on September 11, 1967, a 2-page memo with the subject: "Data Priority Coordination - a plea for help." This "plea for help" memo to the Apollo world was the launching point of his extensive and important mission techniques activity. He requested memo recipients to *"review the format and technical content of the attached Apollo Mission Techniques document prepared by TRW, which, if proper, I would expect will become a prime component of whatever we do."* It was.

Next, on October 19, 1967, Tindall issued four memos that:

(1) established the Ascent/Rendezvous working group and scheduled its first meeting for Oct. 27, 1967, group size - six persons.

This was the first of many subsequent mission techniques meetings.

(2) established the Lunar Descent working group and scheduled its first meeting for Oct 31, 1967, group size - eight persons

(3) transmitted to MSC Systems Engineering Division the TRW Descent phase document - "to be added into the appropriate vacancy in the Apollo Mission Techniques document."

(4) transmitted to MIT the TRW Descent phase document - "to be added into the appropriate vacancy in the Apollo Mission Techniques document I sent you recently."

In these first two memos Tindall requested keeping the working group sizes as small as possible, but that soon gave way to the reality of the magnitude of what needed to be done. For the first six months of 1968 for example, he personally held twelve data priority meetings where average attendance was about 22 persons, including two astronauts and 3 TRW engineers (these are averages). This did not include our smaller working group meetings to develop and refine the logic flow diagrams with decision limits, etc.

Note that for the first meeting - to review the ascent/rendezvous phase document and the logic flow charts - there were only six of us, including Tindall, myself and astronaut Pete Conrad (later the third man to walk on the moon).

This first meeting was also my first time working with one of the astronauts, but there were several others. I particularly remember Buzz Aldrin (later the second man to walk on the moon) because of his enthusiasm and many ideas as to how things could or should be done – a very bright and creative individual.

Part III herein shows Tindall's "a plea for help" memo, the 4 memos listed above and others important to this story. After this launch of Tindall's mission techniques activity, my TRW team went on to produce all the subsequent mission techniques documents, up until the first moon landing flight.

As Chief, Apollo Data Priority Coordination, Tindall was one of the key figures in NASA MSC management. His contributions from the many meetings he chaired were crucial in making sure that communications between the many parts of the Apollo program worked in harmony. His unique ability to get people talking and communicating clearly and openly became vital to the flight and ground crews that were using the vast complex of hardware and software that was the Apollo program. His memos documenting the outcomes of his many meetings became known as "Tindallgrams" – and they became legendary – they were clear, informative and succinct, their tone often folksy and humorous.

I was fortunate to have worked directly under Tindall for two-plus years, leading the TRW team that wrote and published the Apollo Mission Techniques documents for all manned missions - until the first moon landing.

Part III of this book describes this activity in detail, from its early beginnings until the first moon landing flight – Apollo 11.

Apollo 8 – The First Humans To Orbit The Moon (Dec. 1968)

The historic Apollo 8 mission set a new space milestone; astronauts Borman, Lovell and Anders became the first humans to see the backside of the moon and to actually go into orbit around the moon. With their worldwide telecast message on Christmas Eve and that unforgettable photo taken of the earth, 207,000 miles away above the moon's horizon, it was a joyous and high-time in Houston – we were really on the threshold of accomplishing a moon landing! Our lunar landing goal was one big step closer and optimism was high - we had successfully gone to the moon, orbited it for 10 revolutions, and returned our astronauts safely back to earth.

The 7-day mission consisted of a 3-day zero-g coast period from the earth to the moon followed by the Lunar Orbit Insertion (LOI) burn. After 20 hours in lunar orbit (10 orbits around the moon, each with a period of 2 hours) the Trans Earth Insertion (TEI) burn placed the spacecraft onto its second 3-day coast period followed by reentry and splashdown.

During their 20 hours in lunar orbit, the crew conducted a full, sleepless schedule of tasks, including landmark and landing site tracking, stereo photography, and sextant navigation. Six real time telecasts, all of excellent quality, were transmitted worldwide to all five continents. During a telecast on Christmas Eve, the crew read verses from Genesis and wished viewers,

Apollo 8 in orbit around the moon views the earth

"Good night, good luck, a Merry Christmas and God bless all of you - all of you on the good Earth."

The Apollo 8 mission was originally to be flown as a "D" mission type - "Manned CSM and LM operations in low Earth orbit" but the LM would not be ready in time for the planned December 1968 launch date. But the Apollo 8 CSM, the three astronauts and the Saturn V launch vehicle were all ready to go, so a new mission type, "C-Prime" had earlier been defined: "Manned lunar orbit flight with CSM only."

The Apollo 8 decision to commit a crew to a manned *lunar orbit* flight on only the third launch of the Saturn V was a gutsy decision. Although the first Saturn V launch (the unmanned Apollo 4 flight) was a complete success, the second Saturn V launch (the unmanned Apollo 6 flight), experienced major malfunctions caused by significant "pogo" vibration levels. They were not catastrophic, but resulted in J-2 engine early/incorrect shutdowns, which in turn resulted in the CSM not achieving its planned orbit and consequently the planned repeat demonstration of the CM heat shield's capability to withstand a simulated lunar re-entry was unsuccessful. But the heat shield was not the primary concern; the primary concern was that significant levels of pogo vibration could cause structural/catastrophic failure, i.e., would the pogo fixes/work-arounds for this third Saturn V launch vehicle, to be used now for a *manned* mission, be successful? Von Braun was confident they would be, NASA management agreed and declared the Saturn V "man ready".

Other concerns existed as well - how to do the lunar orbit insertion burn safely, on the far side of the moon and out of communications with Houston Mission controllers, and the precision of our orbit determination techniques.

George Mueller (head, Office of Manned Space Flight) in a 1989 oral history workshop stated:[11]

> the decision to "go ahead" for the Apollo 8 lunar mission with the Saturn V was "one of the turning points in the program and one of the best things that we did."

Christopher Kraft (Apollo Mission Flight Director) from that workshop:

> "that (decision) took a lot of guts, a lot of nerve, but I think we knew what we were doing. And we looked at that and said the risk is worth the gain. So let's go and orbit around the Moon.

> Now, there were a lot of white-faced astronauts when we said that. But nevertheless, we again said that we thought that risk was worth the gain. And it was."

Kraft further stated:

> "When George Low asked us what we could do to go around the Moon, we decided look, we are having a hell of a lot of trouble with this lunar orbit determination. We have been looking at the lunar obiter data (the Boeing satellites that were orbiting and mapping the moon), and we can't figure out where the spacecraft is when it comes back around. We are several thousand feet off."

---

[11] From an oral history workshop conducted July 21, 1989 "Managing the Moon Program: Lessons Learned from Project Apollo". Go to: *https://history.nasa.gov/monograph14.pdf*

108

Bill Tindall from that same workshop:

> "We knew what we were doing, we had procedures laid out, we tried to imagine every single conceivable failure that could occur and what we were going to do about them. We did things like lunar orbit insertion in two stages instead of one big burn because an over-burn that might go undetected would cause it to crash into the Moon. So we backed off and did a two-burn."

Two months before the Apollo 8 flight, Apollo 7, the first manned flight in the Apollo program, flew with the redesigned Block II CSM and had "performed superbly" for 10.8 days, longer than needed for a journey to the moon and back. And on that flight the SPS, the rocket engine that *must* work to insert the CSM into and out of lunar orbit, performed "eight nearly perfect firings out of eight attempts."

Note that the Apollo 7 mission used the Saturn-IB launch vehicle which had never experienced a pogo problem.

The Apollo 8 decision was also consistent with NASA Administrator James Webb's direction to prepare Apollo 8 for a possible lunar orbit mission in 1968 as he had been informed by CIA sources in 1968 that the Soviet Union was developing its own heavy N-1 rocket for a manned lunar mission.[12]

But no doubt many breathed a big sigh of relief 12 minutes after liftoff when the S-IVB's first burn ended, having successfully inserted the CSM into a safe earth parking orbit without experiencing pogo oscillations large enough to cause problems in any of the three stages. That left only the second S-IVB burn of 6 minutes to put them onto a translunar trajectory, which was also accomplished successfully.

Following Apollo 8, nine more Saturn V launch vehicles were flown to complete the Apollo program. All were successful, thanks to Wernher von Braun's team of rocket engineers and their many supporting contractors.

---

[12] Revelations after the collapse of the Soviet Union supported the CIA's input to Webb that the Soviets were indeed in 1968 working aggressively towards a moon landing. Go to: https://en.wikipedia.org/wiki/James_E._Webb.

The Astronauts Silver Snoopy Award

The "Silver Snoopy" is awarded to individuals for outstanding efforts that contribute to the success of human space flight missions.

Silver Snoopy
Lapel Pin

The award is a sterling silver pin in the form of the "Snoopy" character wearing a space helmet and space suit. Recipients are also given a letter of commendation personally signed by an astronaut, citing the astronauts' appreciation of their performance.

To meet the criteria for this award, the individual's work must be outstanding to distinguish the individual in his or her particular area of responsibility, and it must make a meaningful contribution to flight safety or mission success.

Two months after the drama and excitement of the Apollo 8 mission, in February 1969, ceremonies were held at TRW Houston Operations where Astronaut James Irwin presented silver "Snoopy" pins to 21 Houston Operations employees and two from Redondo Beach. These employees were recognized for their individual contributions in trajectory design, simulation, analyses and systems engineering support to the NASA/MSC Apollo Spacecraft Program Office.

ASTRONAUT JIM IRWIN paid a surprise visit to TRW's Houston Operations recently to honor the above employees for work excellence in support of the Apollo program. Pictured are, first row, left to right; Richard Boudreau, Bob Hopkins, Patricia Vander Stucken, Karen Christie, John C. Miller, Lt. Col. Irwin; second row; Robert Gerbracht, Thomas Barrie, Hugh Freeman, Gordon, Teveldahl, John Aha; third row; Arnold Rosenbloom, Manager, Houston Operations; Charles Drinnan, Ed Mellon, Arthur Satin, Peter Janak, Owen Bergman, and fourth row; James Dexler, William Lee and John Norton. Other honorees not pictured, were; Amelia Goldenbaum, William M. Lear, Robert H. Manders, Diana L. Shaffer and John Walker.

I was included here for the Apollo mission techniques work.

Many of the awards resulted from significant real-time contributions during the Apollo 8 mission. Owen Bergman, for example, was cited for his assumption – on very short notice – of the responsibility for developing a real time procedure to provide an independent evaluation of the Apollo 8 navigation accuracy. His work confirmed the accuracy of the Real time Computation Center at MSC and spacecraft navigation, thereby increased the level of confidence in the safety of the Apollo 8 mission as the spacecraft approached the Moon.

Following is the letter of commendation I received from Astronaut James B. Irwin for the mission techniques work.

NATIONAL AERONAUTICS AND SPACE ADMINISTRATION
MANNED SPACECRAFT CENTER
HOUSTON, TEXAS 77058

IN REPLY REFER TO: CB   February 3, 1969

Mr. Richard J. Boudreau
Systems Group of TRW, Inc.
Houston Operations
P.O. Box 58327
Houston, Texas 77058

Dear Mr. Boudreau:

As you know, the Apollo Program succeeded in accomplishing two extremely complex manned missions during 1968. These successes would not have been possible without exceptional professional effort and personal dedication to excellence and quality on the part of the entire manned space flight team.

Special recognition must, however, go to those who have put forth far greater effort than called for in the performance of their program responsibilities.

This letter is to express to you my personal appreciation for the care and effort you have demonstrated in performance of your duties on the Apollo Program, and especially for your contributions to the development of Apollo mission techniques that define the guidance and control sequence of events, the data flow and real time decision logic for various phases of the Apollo missions.

As a token of appreciation, please accept the astronauts "Silver Snoopy" award for professional excellence. We hope that you will wear it with deserved pride, knowing that it is given only to those individuals whom we regard as among the best in their respective professions.

Congratulations and best wishes for continued achievement.

                          Sincerely,

                          James B. Irwin

                          NASA Astronaut

Anna's question: Papa, is your name really on the moon?

Anna asked me this question years ago, and my somewhat stumbling response then was, basically, I don't know. Upon further consideration, the short answer to your question is "probably not."

What happened? I received a call from someone in the Astronaut office stating that I had been selected for a Silver Snoopy award, and that my name would be on a plaque to be left on the moon. Other engineers told me they received a similar phone call; apparently I was not the lone ranger here.

At the time I paid this no further attention – not important, busy with work, never heard anything more. Searching online, I was unable to verify whether *any* names are on the moon other than a few on well-known plaques.

Anyway, if my name *is* on the moon, it's surely of microdot-sized text.

So what happened? My guesses at possibilities:

(1) Snoopy awardee names ARE NOT on the moon (most likely)
    a. It was planned to be done, the calls were made, but it was never finally approved or later the decision was rescinded.
    b. mistaken calls based on misinformation or a misinterpretation
    c. prank calls by a jokester

(2) Snoopy awardee names ARE on the moon
    a. it was *planned* to be done, calls were made, then later rescinded, but someone (the astronauts *might* do this) went ahead and did it anyway
    b. it was *planned* to be done, calls were made, and the names *were* placed on the moon, but no evidence can be found that this happened for some unknown reason.

So Anna, my best guess; my name is almost certainly not on the moon. But who knows, it might be there, what one could call a known unknown.

Note:

In February 2002, Donald Rumsfeld, the then US Secretary of State for Defense, stated at a Defense Department briefing: "There are known knowns. These are things we know that we know. There are known unknowns. That is to say, these are things that we now know we don't know. But there are also unknown unknowns. These are things we do not know we don't know."

As a result, he was almost universally ridiculed; many thought the statement was nonsense. However, careful examination of the statement reveals that it does make sense, indeed the concept of the unknown unknowns existed long before Donald Rumsfeld gave it a new audience.

IT WAS NOT ALL WORK RELATED

Two years after starting work in Houston, our son Roland was born. Michele was 3 then, with her 4th birthday party coming up in December. We were in Houston for 9 years before I transferred to TRW's main facility in Redondo Beach, California. So as the kids grew up in Houston there were violin lesson for Roland starting at 3, then progressing to larger and larger instruments as he grew and piano lessons for Michele. For the many camping trips with our Brabant neighbors, we had all the gear, tents, sleeping bags and cots, kerosene lights, scorpions too - all included.

I joined the TRW golf club and will never forget that first match in Galveston, trying unsuccessfully to avoid "casual" water (where you could move the ball without penalty), but there was only casual land, and standing on ant hills without protective spray.

Wilma and I both joined TRW's bowling league and won trophies – we were not the best, just most improved in the handicap section of the league.

Clear Lake was right next door to Clear Lake City where we lived, so I bought a Dolphin Sr. sailboat and joined the Clear Lake Sailing Club to participate in their class-boat races. A huge diversion from work - competing with identical sailboats one had to live totally in the present, focused only on the wind direction, controlling the boat's sail and rudder, your position on the boat, and where the other boats are relative to you and the finish line - Great fun! We also of course took the boat with us when going to the beach with Wilma and kids. Once when sailing with Michele on Galveston Bay I karate-chopped a small shark that was surfing on the boat's wake to chase it away.

My guitar interest started in Houston - I studied Flamenco guitar playing and took folk guitar lessons by Laura Weber via the Public TV channel, ended up with three guitars, one bought in Madrid and another in Monterrey, Mexico, when we went to Laredo, Texas, to buy tortillas for a Mexican party Wilma was hosting in our home, with Bill Bennett playing great flamenco guitar.

Exciting times during our 9 years in Houston – many stories yet to write. Our monthly bridge parties that ended up being gourmet dinners, cockroaches and cottonmouths in our garage, Wilma's mom lost on her way from LA, Leo's kids living with us, Thanksgivings with Pat & George, Kathleen's marriage to Tommy Cox and her son Brian born - all part of our Houston adventure.

And we attended moon landing celebrations as they occurred and as we were invited. Shortly after the first moon landing flight, we got tickets for "Houston's Salute to NASA and Apollo 11" attended by Frank Sinatra, the three Apollo 11 astronauts and their wives and many others.

Later on July 22, 1972, we attended the "Moonwaltz Ball" at River Oaks Country Club where TRW and others were sponsors. This party commemorated the landing of Christopher Columbus in the New Word in 1492 and the first lunar landing July 20, 1969. Again there were astronauts with their wives, heart surgeon Dr. Denton Cooley and other important people.

AT OUR HOME IN CLEAR LAKE CITY – NEAR MSC

My wife Wilma with Roland and Michele - 1967

Wilma, Michele and Kathleen - in our mostly unfurnished living room

Roland, 1968

In our front yard

**Dolphin Sr. sailboat –
near Nassau Bay homes**

**Our "Funny Four" team
won a trophy in 1968**

ABC

League Name	TRW SYSTEMS MIXED FOUR "SUMMER" LEAGUE
League President	R. E. Eick Phone HU-8-2605
League Secretary	A. Faye Moore Phone 946-5380
Ass'n	Week of 27 June 1968

TEAM STANDINGS

	NAME	WON	LOST	PERCENTAGE	TOTAL PINS	AVERAGE
1	# 1 - FUNNY FOUR	11	5	.688	6842	570
2	# 10 - TEN PINS	11	5	.688	6234	519
3	# 11 - GOOD GUYS	10	6	.625	6531	544
4	# 2 - NONAMES	10	6	.625	6333	527
5	# 12 - WHITE HATS	9	7	.563	6427	535
6	# 8 - HELLWITHIT	8	8	.500	7083	590
7	# 5 - THIRST SLAKERS	8	8	.500	6810	567
8	# 3 - SPAZTIKS	7	9	.438	6852	571
9	# 7 - ALLEY CATS	7	9	.438	6589	549
10	# 4 - ALLEY OOPS	6	10	.375	5495	457
11	# 9 - REBELS	5	11	.313	6403	533
12	# 6 - SOKITUMY	4	12	.250	5912	492
13						
14						
15						
16						

1st High Team, 3-Games (Hdcp) TEAM # 8 - ALLEY CATS 856   1st High. Team Game (Hdcp) TEAM #1 - FUNNY FOUR 2361

1st High, Ind. 3-Games				1st High, Ind. Game		
W. Scratch	- Joan Eick	512		W. Scratch	- Judy Klenk	207
W. Hdcp.	- Cora Blaylock	626		W. Hdcp.	- Faye Moore	234
M. Scratch	- Tom Fujawa	552		M. Scratch	- R. VanderStucken	216
M. Hdcp.	- Harry O'Haver	648		M. Hdcp.	- E. Herring	248
	R. Boudreau	648			John DeVillier	248

Amount in League Account $_____ Where Deposited _____

DATE PRESIDENT Verified Account _____

Signature of League President _____

INDIVIDUAL AVERAGES

NAME	TOTAL PINS	GAMES	AVERAGE	NAME	TOTAL PINS	GAMES	AVERAGE
TEAM # 1 - FUNNY FOUR				TEAM # 7 - ALLEY CATS			
32 R. VanderStucken ***	1418	9	157	45 Bill Klenk ***	1265	9	140
72 Pat VanderStucken ***	933	9	103	46 Judy Klenk	1660	12	138
46 Wilma Boudreau	1660	12	138	63 Ann Widdifield	1399	12	116
27 Dick Boudreau	1956	12	163	39 Harry Widdifield	1771	12	147
TEAM # 2 - NONAMES							
32 Ron Wendt ***	1885	12	157				
66 S. Wendt	1352	12	112				

READOUT   NOVEMBER 30, 1971

**Wilma-TRW Wives
Brunch, 1971**

APOLLO 11 Salute

THE ASTRODOME
Houston's Salute
to N. A. S. A. and
APOLLO 11

SATURDAY AUG 16 7:30 P. M.

SATURDAY - 7:30 P.M.
No Refunds - No Exchanges

THE PLANNING COMMITTEE for the recent TRW Wives Brunch included (L-R)
Mary Ellen Sievers, program; Sue Huggins, chairman; Ginger Avvenire,
site; Pam Reinhardt, publicity; Sandra Lee, site; Wilma Boudreau, reser-
vations, Lucia Serrell, decorations; Kathy Panneton, reservations; Mary
Lou Russel, program. Dorothy Snygg, decorations, is not pictured.
Mary Dale Swan of Swan Galleries presented a program on the arts which
included a display of hand-crafted items. The Brunch was held at the
Watergate Yacht Club.

PART III
APOLLO MISSION TECHNIQUES
A TRW STORY

"The program office [ASPO] recognized that, and I don't remember the man's name that really surfaced it, but he did the program a favor, and they invented the Apollo Data Priority, which was the precursor name for the mission techniques activity."

Phil Shaffer – Flight Dynamics Manager

Phil replaced Bill Tindall as ASPO/Chief, Apollo Data Priority Coordination – shortly after the Apollo 11 mission.

TRW's Marvin Fox is the man who "did the program a favor" by inventing Apollo Data Priority.

INTRODUCTION

My intent in writing this is to fill a space history void by providing a more complete and accurate description of the Apollo Mission Techniques effort; its early beginnings - why and how this activity was started, who the key players were, the specific events and products produced, and TRW's role in all of this.

My search at NASA and other internet portals led me to the conclusion that such an overview was not to be found. I did find mention of Data Priority/Mission Techniques in many places, many references to Bill Tindall's activities including his "Tindallgrams" and his "Techniques of Controlling the Trajectory" paper, but no complete overview of the significant effort Mr. Tindall managed; its products, key milestones and TRW's intimate involvement with that effort. I also found no mention either of the earlier TRW activities under Dr. Shea that launched the pre-Tindall effort which led to the first Apollo Mission Techniques documents - before the Apollo fire and before George Low became ASPO manager.

Key NASA/MSC participants in the Apollo Program and others agree that the Mission Techniques effort was essential to the success of the Apollo moon landing program. Participant statements reflecting this are included herein as excerpts from the NASA/JSC Oral histories project.

What are "Apollo Mission Techniques?"

The overarching goal of the Apollo Mission Techniques effort was to coordinate all MSC internal and external contractor efforts in developing the procedures for the operational utilization of the trajectory control systems used in all of the manned Apollo missions. This includes the utilization of spacecraft propulsion, guidance and control systems as well as the associated support from the MSFN and the Mission Control Center.

"Decision logic" and "data priority" were basic features of the documented techniques; documented to provide a recognized defined baseline plan for decisions by flight and ground crew personnel. They specified the priority of the data to be used in making decisions that verify whether or not the GNC systems were operating correctly, and identified appropriate actions for non-nominal situations to assure crew safety and, as possible, mission success.

Specific goals of this activity were to:

1. determine the operational rules and procedures for properly using the Apollo systems, including primary and backup systems;

2. investigate the systems capabilities and constraints and evaluate their accuracies; establish the criteria for system selection during various phases of the mission;

3. establish the proper spacecraft and ground displays and use of these displays.

4. provide specific guidance for ongoing spacecraft hardware development by NAR, GAEC and others, onboard software development by MIT, ground software development by many, mission planning, astronaut flight plans and procedures development, and ground-crew plans, procedures and overall mission rules to be developed by MSC and support contractors.

The effort to accomplish these goals produced a large number of Mission Techniques documents that describe a formally approved and controlled baseline for how each Apollo manned mission was to be flown. Documents were produced for each phase of the seven manned Apollo mission types, C, C-Prime, D, F, G, H & J. There were 12 Mission Techniques documents that governed mission type G, the first lunar landing mission flown by Apollo 11. Apollo missions 12 thru 17 were mission types H & J.

Each document describes, for a particular mission phase, the sequence of major guidance, navigation and control (GNC) events, GNC data flow/transfers between the spacecraft and ground, and most key, the rules and decision logic to be used in real-time (during the flight) by the astronauts and ground controllers to determine the health of the onboard GNC systems and to accomplish proper targeting and control of the spacecraft rocket engine burns to change the spacecraft trajectory as needed.

Most of this information is presented in long flow charts at the end of each document. These charts present the time sequence of the real-time activities and decision logic for that documents mission phase – the steps to be taken under each condition in monitoring the guidance and control systems performance, onboard and on the ground and the procedures to be followed if certain limits are exceeded.

The logic flow charts were the results of the many "data priority" meetings, "Tindallgrams" and mission phase working group meetings - many man-hours of effort to "get it right." These charts are not meant to be followed step by step, but are rather a guide for the developers of ground and flight crew procedures, as well as providing guidance to flight controllers and crew during simulated and real mission operations.

The mission techniques documents were considered controlled documents, and had precedence over Mission Rules, the Apollo Operations Handbook, flight crew procedures and flight plans. They influenced many other mission planning activities and had a large distribution list to assure a clear understanding and agreement among the many involved MSC and contractor parties.

W. David Woods' excellent book *How Apollo Flew to the Moon* provides astronaut David Scott's perspective on mission techniques:

> ***Mission Techniques.*** *These defined the manner in which the mission would be flown, or more specifically, the manner in which the spacecraft "trajectory" would be controlled. Once the mission objectives, the crew procedures, and the trajectory are defined, it becomes necessary to define how the various components of the guidance, navigation and control systems, as well as the rocket engines, should be used during each phase of the mission to maintain the "trajectory" for that phase. The mission techniques development task was basically: how do you decide how to fly an Apollo mission? It required detailed planning on precisely how well the systems must work to achieve the mission, including all of the options for use and/or failure of primary or backup systems. This was also termed*

the "data priority" task, and its detailed planning was absolutely essential for mission success.

As an example, LOR was key to both mission success and crew safety. During the LOR phase. five first-class systems computed the rendezvous manoeuvers – two in the LM, one in the CSM the MCC, and even some simple charts used by the crew. But if there was disagreement, there had to be a rationale for deciding which one to use.`

Mission rules were established as a combination of crew procedures and mission techniques whereby, if any failure or anomaly occurred, a "rule" defined the action to take. Mission rules essentially answered the "what if" question. Even so, there were events during Apollo that had not been foreseen, and required thinking and action beyond mission rules – an excellent example being the oxygen tank explosion that aborted the Apollo 13 mission.

Mr. Woods' book is a comprehensive look at the equipment, procedures and techniques that, as he states, describes "the practical aspects of how the voyage from the Earth to the Moon was achieved." It provides a detailed descriptions of the many GNC system components and how they are used during each mission phase. Also included is a mostly-GNC glossary (even a definition of REFSMMAT is included for example), and in an Appendix, a brief functional description of each of the 76 computer programs residing in the CMC and LGC.

This is a book well worth reading; especially interesting are the descriptions of individual astronaut experiences during some of the more time-critical periods in their missions.

Trajectory Control Systems and Operational Responsibilities

The major GNC systems involved with trajectory control of the two Apollo spacecraft included: (1) the ground-based Mission Control Center-Houston (MCC-H) - ground operations personnel supported by the Manned Space Flight Tracking Network (MSFN) and the Real-Time Computing Complex (RTCC); and (2) all the onboard astronaut-controlled GNC systems - a PGNCS in both the Command Module (CM) and the LM, optical sighting devices for IMU alignments and drift checks, LM landing and rendezvous radars, and PGNCS backup systems – the AGS in the LM, and the Stabilization & Control System (SCS) and Entry Monitoring System (EMS) in the CM.

MCC-H had prime responsibility for spacecraft tracking and navigation and providing state vector, REFSMMAT, ΔV and other uploads to the spacecraft GNC systems and crew, as well as monitoring GNC systems health and functionality. The astronauts and onboard GNC systems were responsible for onboard GNC systems management including star sightings etc. to provide information needed by the ground, and for spacecraft navigation, guidance and control during the ΔV ("delta V") maneuvers - incremental velocity changes in the trajectory via rocket engine "burns".

A key part of the mission techniques effort included systems and trajectory analyses by MSC and TRW to determine *nominal and expected* deviations in spacecraft trajectory and propellant usage due to expected GNC system errors (IMU drifts and biases, etc.) as well as expected MSFN tracking errors. *Allowable* deviations from these expected values were also estimated, these dependent upon the particular real-time segment of the mission. If deviations observed in real-time by ground crew and/or astronauts were greater than expected, these would be indicative of a degraded or failure situation requiring real-time decisions for corrective or abort actions.

These analyses were especially important for critical mission events such as the ΔV burns needed to attain a safe lunar orbit, lunar descent/landing at the desired site, LM/CSM rendezvous and safe earth return/reentry trajectories.

Because of the many real-time sources of GNC data, "Data Priority" was a key aspect of the techniques development, and Bill Tindall, as NASA's "Chief of Apollo Data Priority Coordination" was the very effective leader of this two-plus year effort - until the first moon landing by Apollo 11.

EARLY MISSION TECHNIQUES DEVELOPMENT

In early 1966 TRW had initiated a "Data Priority" effort for ASPO to develop and produce Apollo Mission Techniques. This early work led to the development of the first two mission techniques documents – the first for Lunar Ascent/Rendezvous and a month later the second for Lunar Descent, the two most critical mission phases.

After the Apollo fire in January 1967, Howard W. "Bill" Tindall Jr. was appointed Chief, Apollo Data Priority Coordination and these first two documents were the starting point of his continuing efforts in this over the next 2 years - until the first lunar landing.

I was the lead engineer for TRW's initial efforts under Dr. Shea, then continued as manager/performer for TRW's efforts for the next 2 years essentially as staff support to Bill Tindall. My TRW team, by participating in the various working groups established by Mr. Tindall, authored and published all the Apollo Mission Techniques documents produced from these working group meetings.

As the FORWARD in every mission techniques document states:

"For each mission phase, a Data Priority Working Group has been established under the direction of the Chief, Apollo Data Priority Coordination, ASPO. These groups, which are comprised of representatives of MSC and support contractors, hold frequent meetings to coordinate their various associated activities and develop agreed upon mission techniques. TRW Systems Group assists in the development of the techniques and documents them for ASPO. After formal review by ASPO, E&DD, FCOD, FOD, GAEC, MDC, NR and TRW, a document such as this one is issued."

The early work by TRW – Marvin Fox

TRW identified the need for a data priority effort back in early 1966 when Dr. Joseph Shea was ASPO Manager. Dr. Shea initiated an approximate two-man task first reporting directly to himself and then later to Richard Carley who worked for Shea. Dr. Arnie Rosenbloom (Houston Operations manager at that time), Marvin Fox (my TRW supervisor) and I were the initial performers on this task.

Marvin Fox – 1967

It all started with Gemini – the 2-man capsule flown in earth orbit to develop and validate the critical rendezvous and docking operations that would be needed for the Apollo missions. Data Priority issues were identified by TRW's post-flight analyses of Gemini flight data. These analyses showed that the astronauts were not burning a single ΔV solution for their rendezvous burns – that is, they would select the components from either crew onboard chart solutions, the on-board G&N, or MSFN (ground) supplied solutions, or some sort of average of all of these. There were no specific procedures, data acceptance limits, and decision logic for the ΔV burns – a "data priority" problem. For Gemini (especially Gemini 10), this led to excess propellant usage, and could have led to a failed rendezvous and/or unsafe reentry trajectories.

Early in 1966 Marvin had recognized that because of the many sources of Apollo GNC data and the complexity and extent of the Apollo mission timelines, there was a critical need to have a documented and controlled baseline plan for trajectory control for all Apollo manned missions, including the real-time decision logic used by the onboard crew (astronauts) and ground flight controllers for GNC data verification, ΔV maneuver determination, and actions to be taken when non-nominal situations arose. He called this Data Priority, and enthusiastically promoted the need for this effort to key NASA/MSC individuals within ASPO and other MSC organizations.

Shown here is a briefing given on September 10, 1966 to ASPO/Dr. Shea, MPAD, GCD and FCSD. This briefing contained detailed results and analysis of Gemini rendezvous maneuvers with observations and conclusions applicable to Apollo. This work was done under the technical direction and leadership of Mr. Fox.

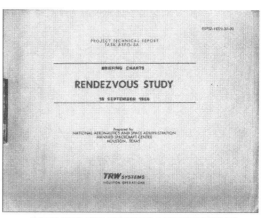

I donated the complete set of these briefing charts to the University of Houston Clear Lake (UHCL) Archives.

EARLY TRW STUDY ADDRESSING GEMINI DATA PRIORITY ISSUES AND THEIR APPLICABILITY TO APOLLO

Another study was done in 1966 by a four man G&C Division panel that recommended redesign of the Flight Director Attitude Indicator (FDAI) display electronics so that instead of the "eight-ball" display only showing spacecraft attitude referenced to a pre-defined inertial coordinate frame, the crew could also select the display to show spacecraft pitch attitude referenced to the ground below, or a "local vertical" reference (local vertical is a line from the orbiting spacecraft to the center of the earth or moon). Pilots are familiar with this type of "ground below" FDAI display reference and it was used successfully in the Gemini spacecraft for rendezvous operations.

ORDEAL capability as envisioned would consist of fairly simple electronics to drive the FDAI eight-ball pitch attitude down at an astronaut-selectable constant orbital rate to provide the local vertical reference – approximately 360 degrees per 90 minutes when in earth orbit and 360 degrees per 2 hours for lunar orbits. It takes about 90 minutes to orbit the earth and 2 hours to orbit the moon - for the typical low altitude orbits used in Apollo. As implemented, this was called ORDEAL for Orbital Rate Drive Electronics for Apollo and LM.

NAA and GAEC were then directed to supply the ORDEAL capability, but after receiving their cost proposals, Dr. Shea stopped their ORDEAL work - "in the interests of fiscal economy" as shown in his September 26, 1966 memo on the following page.

It was later decided to provide the units to NAA and GAEC as Government Furnished Equipment (GFE) due to the fact that the units were essentially identical. This was done and according to an Apollo Experience Report the ORDEAL GFE was flown with no anomalies on all manned missions - 11 Command Modules and 9 Lunar Modules, and "The decision to supply the orbital-rate-drive assembly as Government-furnished equipment resulted in a significant cost saving for the Government."

Dr. Shea in this memo also directed that the Guidance and Control Division develop G&C guidelines coordinated with all relevant MSC organizations and TRW – who had previously "done work related to coordinated crew procedures, software output to displays, operational rules such as procedures for selection of values of ΔV at burn points from the diverse informational sources available."

OPTIONAL FORM NO. 10
MAY 1962 EDITION
GSA FPMR (41 CFR) 101-11.6

UNITED STATES GOVERNMENT

Memorandum

Logged
OCT 3 1966
ap

TO : EG/Chief, Guidance and Control Division

DATE: September 27, 1966

FROM : PA/Manager, Apollo Spacecraft Program Office

SUBJECT: Guidance and Control Guidelines for Missions AS-278, AS-503 and AS-504

It is requested that you perform a comprehensive study to establish the G&C guidelines for Missions 278, 503 and 504. This study is to include the establishment of crew procedures, detailed software outputs to displays, operational rules such as procedures for selection of values of ΔV at burn points from the diverse informational sources available, G&C monitor procedures by the crew and on the ground by Flight Control personnel, etc. I want to assure that mission analysis, flight planning, hardware constraints, computer programming and crew training for Apollo are carefully coordinated so that all facets of the Apollo rendezvous and lunar missions are compatible from technical and schedule standpoints. I consider that this is the responsibility of the G&C Division and that you are to be supported by, and are to solicit the help of, all other elements of MSC which are doing work in some of these areas. Specifically, MPAD, FCSD, IESD, Astronaut Office, ASPO-MOD, and TRW have done work related to this problem and are requested to coordinate their work through you. You are requested to review their work, utilize it as appropriate, and redirect it as necessary to assure that all G&C hardware and operational objectives of Apollo are met on time.

In the interest of fiscal economy, I have verbally directed that all funded work on ORDEAL (orbital rate FDAI) be stopped at NAA and GAEC. Although an orbital rate FDAI was used in Gemini, I have not been convinced in discussions with you and with personnel from FCOD that such a display is required to fulfill Apollo objectives.

I would like to receive from you by October 31, 1966 a detailed report of "G&C Guidelines" for missions 278 and 504. I would like to receive from you by October 17, 1966 the "G&C Guidelines" for the rendezvous portions of 278, 503 and 504. At that time, I would like your specific recommendations concerning the necessity of incorporating ORDEAL into the CM and LM in order to perform these missions. Because of the hardware, schedule and fiscal impact of ORDEAL, I do not want to incorporate it into the spacecraft unless there are strong arguments shown that the missions cannot be performed with the current hardware configuration.

Buy U.S. Savings Bonds Regularly on the Payroll Savings Plan

You are also requested to assure that the baseline software displays from both the DSKY and DEDA are consistent with the baseline hardware configuration now approved.

Original Signed By:
Joseph F. Shea

Joseph F. Shea

cc:
AA/ R. R. Gilruth
AB/ G. M. Low
CA/D. K. Slayton
CB/A. B. Shepherd
CB/J. A. McDivitt
CB/ T. P. Stafford
CF/ W. J. North
CF24/P. Kramer
CF32/D. F. Grimm
EA/ M. A. Faget
EA2/J. B. Lee
FA/C. C. Kraft
FA/ S. Sjoberg
FA3/ R. G. Rose
FC/ J. D. Hodge
FM/ J. P. Mayer
FM/ H. W. Tindall
FM/ M. V Jenkins
FM2/L. Dunseith
FM2/T. F. Gibson
FM2/P. J. Stull
PA/W. A. Lee
PA/A. D. Mardel
PE/O. G. Morris
PF/R. W. Lanzkron
PM/O. E. Maynard
PP/J. T. Markley
EG2/D. C. Cheatham
EG4/R. A. Gardiner
EG23/K.J.Cox
EG25/T.V.Chambers
EG27/D. W. Gilbert
EG42/G. Rice
EG43/H. Croyts
EG44/W. J. Rhine

TRW/Rosenbloom
NASA Hqs - Gen. Phillips
        Paul Schrock
        Bellcomm/Thompson

The Very First Mission Techniques Documents

TRW's early Mission Techniques work was focused on those mission phases deemed most critical to Apollo mission success, Lunar Ascent/Rendezvous and Lunar Descent. This work resulted in the creation of two Mission Techniques documents. The first document, dated September 12, 1967, contained an introduction and completed sections for the Lunar Ascent and Rendezvous mission phases. It also contained placeholder tabs for the other mission phase chapters.

The Introduction section covers presentation objectives for documenting vehicle and trajectory control techniques and ground rules to be used for development of the mission techniques and the mission phase operational flow charts.

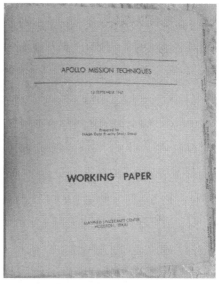

One month later, the second document, dated October 16, 1967, was issued and covered the Lunar Descent mission phase.

The information in these two documents was obtained from MSC personnel and early TRW analytic and simulation studies, and reflects the Data Priority Study Group's work as of August 1967. Each document contains operational flow

THE FIRST APOLLO MISSION TECHNIQUES DOCUMENT – LUNAR ASCENT, RENDEZVOUS

diagrams (on long foldout pages) showing the onboard and ground GNC activities, data flow between the ground and onboard GNC systems, and onboard/ground crew decision logic and with supporting text.

Both were printed and distributed under Tindall's direction shortly after he was appointed Chief, Apollo Data Priority Coordination. The first document was distributed along with his "a plea for help" memo. Both were reviewed and updated within Tindall's working groups, and were the starting point for his early data priority activities.

132

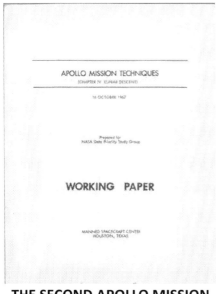

THE SECOND APOLLO MISSION TECHNIQUES DOCUMENT – LUNAR DESCENT

I retained from my Apollo days a paper copy of each of these two fairly large documents (about 330 pages in total) and donated both to the UHCL Archives. Before sending them to UHCL, I scanned the complete first document dated September 12, 1967 (including those long operational flow charts) which I have available in PDF format. I did not scan the Lunar Descent document.

Other important activities resulted from this early Mission Techniques effort - for example the well-known "Lear Filter" developed by Dr. Lear of TRW. This filter resulted from the mission techniques effort in determining how to decide if a PGNCS to AGS switchover is required during Lunar Powered Ascent or during an abort from Lunar Descent. Originally the third vote was to come from MSFN Doppler data compared to PGNCS and AGS derived range data. Our studies showed that this approach was not sensitive enough to out-of-plane guidance errors, so we investigated other techniques, such as using the CSM sextant, LM rendezvous radar data, a roll maneuver after PGNCS/AGS divergence was noted, etc. A Lear filter ultimately became the selected solution for this problem.

Another "Lear" powered-flight data processor was used to accomplish Apollo 12's "pinpoint" landing as described in "Trajectories for the Early Lunar Landing Flights."

Bill Tindall takes over

On January 27, 1967, a routine prelaunch test turned fatal when fire ripped through the Apollo spacecraft cabin, killing all three astronauts. This test was in preparation for the first manned test flight of the Apollo CSM in Earth orbit; Apollo 7. Three months after the fire, in April 1967, George Low replaced Dr. Shea as the new manager of ASPO; the Apollo 7 flight didn't take place until October 11, 1968, a delay of almost 2 years.

Shortly before the fire, Dr. Shea had appointed Richard Carley as the MSC leader of the Data Priority Working Group, providing direction to TRW and MSC support persons.

On May 5, 1967, Mr. Carley sent a 3-page memo to George Low describing the critical importance of developing Apollo Mission Techniques. He recommended that ASPO "implement a common concept of the Apollo mission" stating that "the existing data priority study with TRW can be used to accomplish this objective." He also made recommendations regarding the responsibilities of the various MSC directorates, NAA and GAEC, including "appropriate letters of direction establishing the data priority activities, change control activities on the techniques document, procedures development, and training activity."

Four months later, in August 1967, George Low appointed Howard W. "Bill" Tindall as Chief, Apollo Data Priority Coordination, responsible to "coordinate all MSC and contractor efforts in developing the techniques and procedures for the operational utilization of the trajectory control systems involved in manned Apollo missions."

Bill Tindall

Following Richard Carley's memo is George Low's announcement of Mr. Tindall's appointment.

Two months later Bill launched his extensive Apollo Mission Techniques/Data Priority effort with his "a plea for help" memo issued on September 11, 1967. This memo discusses Data Priority objectives, mentioned the earlier TRW work for Dr. Shea and Dick Carley, and had attached to it - for review - the first published Apollo Mission Techniques document. It also covered Mr. Tindall's proposed plans for continuing mission techniques development and for their use in controlling other relevant MSC activities to be mutually consistent with an established baseline - an excellent summary of previous and planned Data Priority efforts.

134

Development of Mission Operating Techniques - Data Priority and
Procedures Development

Development of adequate system operations techniques and procedures
for the Apollo vehicles especially in the broad G&C area is a critical
factor in the success of the Apollo Program. The vehicles involve
many contractors equipment including GFE items. The development of
proper operating techniques and procedures requires assimulation of
large quantities of data generated to support the hardware design
and integrating this together with the ground systems capabilities to
determine the most suitable techniques and procedures. Undue weight
on using either the onboard on the ground data or equipment or chart
derived data can seriously compromise the mission success. The development of these techniques is equally or more critical to mission success
than the selection and development of specific hardware characteristics.
And obviously selection of hardware characteristics depends on a thorough
understanding of the operating techniques or on an assumption on what
techniques will be adopted. Since many organizations both internal to
and external to NASA have been forced to make these kinds of assumptions
in order to proceed with their particular responsibility it is especially
necessary at this point in the program to review and summarize these
assumptions. Experience of aircraft flight programs, Mercury and Gemini,
has shown that threats to mission success are ~~especially or~~ more likely
to occur due to procedures related activities than to specific hardware
failures. In fact, minor hardware malfunctions can and have been parlayed
into major mission failures unnecessarily due to inadequate or misunderstood procedure related activities.

It is proposed that a plan be developed by ASPO that properly informs
management and that management agrees, adopts, and implements a common
concept of the Apollo mission. The existing data priority study with
TRW can be used to accomplish this objective. Figure 1 shows a functional
block diagram, relating the data priority activities to the Change Board
and the Center and NAA, GAEC and MIT activities. Figure 2 shows the
Flight Operations directorates relationships to the proposed management
plan. Figure 3 shows the Flight Operations directorate relationship to
the Change Board and the Center and NAA, GAEC and MIT activities. Figure
4 shows the engineering directorate relationship to the Change Board and
the Center and NAA, GAEC and MIT activities.

The objectives of the management scheme are to identify that portion of the procedures development called techniques to be identified and agreed to by management. Procedures would then be developed from these techniques. In order to assure maximum compatibility of the developed hardware and the detailed procedures and in view of the fact that the contractors possess the most realistic simulations, it is proposed that the NAA and GAEC engineering simulators be developed into the reference simulations for the Apollo Program; and that NAA and GAEC be responsible for developing and testing the validity and demonstrate to NASA the procedures to be used on each mission as developed from the approved techniques. The techniques would be described in a controlled document, changes to which would be approved by the Apollo Change Board. This techniques document would be organized with chapters for each mission phase. Each chapter would include a description of the mission phase, logic diagrams showing the uses of the onboard primary, secondary systems and charts, and the ground RTCC and auxiliary computer data for G&C decisions. Appropriate figures and graphs show the relative accuracies of all the onboard and ground data, and the third element of the document. This techniques document would be developed first for the lunar mission, then for each manned mission to be flown and developed and approved by management on a schedule that will permit adequate time for the development of the detailed procedures by the contractors; and show three months of time after the procedures development is completed for mission training on the mission simulators at NASA using these procedures. By having the contractors develop the procedures he can be made responsible for a suitable marriage between procedures and hardware and therefore his development in each of these areas can be for mission success criteria rather than to detailed NASA arbitrary requirements.

It is proposed that the engineering directorate be responsible to ASPO for establishing the configuration of all simulators built using Apollo funds and to report schedules, reference letter PA/C67-5 enclosed, and to control the content of engineering memorandum to be issued after each mission phase of each simulation activities (hardware, software, procedures development or crew familiarization).

The Flight Crew Operations directorate would be responsible for monitoring the contractors by being in attendance at the facility to assure that the use of these facilities to develop from the techniques document adequate mission procedures for each mission phase. Flight Crew Operations Division will be responsible to implement training in accordance with these procedures.

The software development activities of ASPO and FOD shown in Figure 5 would be responsible for establishing capabilities to support the techniques document and to support and to submit change proposals and to assure their adequacy in the area identified by the document.

The Mission Operations Division of ASPO would be responsible to identify
the personnel to be present during procedures development and for the
integration of the procedures development from the techniques document.
Obtaining adequate signoffs and assuring that adequate familiarization
has occurred and necessary conditions are met prior to the start of
each phase at each simulation activity.

A preliminary copy to be used as a guide for the development of subsequent chapters of the techniques document is enclosed. Section
of the document, page   identifies conclusions reached and actions
required to allow implementation of the techniques proposed in the
document for each mission phase. In the approved document these
sections would be removed and converted into CCB actions or ASPO actions
as appropriate. The development of the techniques document will allow
the identification of areas not previously analyzed for assignment by
ASPO to Center elements and assure generation and accumulation of
relevant technical data for the establishment of proper operating techniques.

Recommendations

(a) It is requested that the management scheme proposed be discussed
with the appropriate directorates and be implemented with appropriate
letters of direction establishing the data priority activities, change
control activities on the techniques document, procedures development,
and training activity.

(b) For similar reasons to those discussed regarding GSC, it is
recommended that NAA and GAEC be made responsible for the development
of all systems procedures and not just supply data to the best of their
judgment (dependent on personnel used) to NASA for NASA development of
and integration of procedures. The contractors should be responsible
for developing final procedures and all required changes to procedures
should be made by the contractor and the contractor should be responsible
for certifying the procedures by adequate system demonstrations for
normal and failed equipment operation. NASA personnel will be able to
change the contractor response through CCB activity.

Richard R. Carley

Enclosure

No.: 67-114

Date: August 1, 1967

File:

MANNED SPACECRAFT CENTER

ANNOUNCEMENT

CHIEF OF APOLLO DATA PRIORITY COORDINATION

APOLLO SPACECRAFT PROGRAM OFFICE

Mr. Howard W. Tindall, Jr., has been appointed Chief of Apollo Data Priority Coordination. In this assignment he will coordinate all MSC and contractor efforts in developing the techniques and procedures for the operational utilization of the trajectory-control systems involved in manned Apollo missions. This effort has in the past been called "G&N Operations" or "Data Priority specifications." Specifically, his job is to determine the operational rules and procedures for properly utilizing the Apollo systems, including primary and backup systems; to investigate the system capabilities and constraints and to evaluate their accuracies; to establish the criteria for system selection during various phases of the mission; and to establish the proper spacecraft and ground displays and use of these displays. His responsibility will include the utilization of the spacecraft propulsion, guidance and control systems, as well as the associated support from the MSFN and Mission Control Center.

Mr. Tindall will report directly to the Manager, Apollo Spacecraft Program and is, in effect, on loan from his present position as Deputy Chief of the Mission Planning and Analysis Division in the Flight Operations Directorate. A charter delineating his responsibilities and mode of operation will be distributed within the next several weeks. I request that you give Mr. Tindall the fullest cooperation in this important effort.

George M. Low
Manager
Apollo Spacecraft Program

DISTRIBUTION:
Y

OPTIONAL FORM NO. 10
MAY 1962 EDITION
GSA FPMR (41 CFR) 101-11.6

UNITED STATES GOVERNMENT

Memorandum

TO : See list below

DATE: SEP 11 1967
67-PA-T-79A

FROM : PA/Chief, Apollo Data Priority Coordination

SUBJECT: Data Priority Coordination - a plea for help

1. Over the past year considerable thought has been given to how to coordinate mission planning and procedures development involving use of the various guidance and control systems for the lunar landing mission. Initially, a four man panel (Cohen, Jenkins, Smith and Kramer) worked for a month or so and assembled a pretty good preliminary set of logic flow charts describing spacecraft and ground activity starting in the nominal mission with the LM on the lunar surface and ending with rendezvous. This work, you recall, ended with a recommendation to implement the orbit rate ball in the spacecraft and the panel was disbanded. Subsequently, a task was assigned TRW-Houston to carry on this work for the entire lunar landing mission, first under the direction of Dr. Shea and then Dick Carley. In my new assignment as Chief, Apollo Data Priority Coordination, I have inherited this imposing task. This memorandum is to request your assistance in two respects. The first is to solicit your opinion on a proposed approach to the overall coordination problem. The second is to request your review of the format and technical content of the attached Apollo Mission Techniques Document prepared by TRW, which, if proper, I would expect will become a prime component of whatever we do.

2. It has been proposed that the techniques to be used on Apollo missions should be thoroughly documented and maintained under configuration control. Among other things the intent is to make sure everyone working on Apollo who has interest in this business would know exactly what the officially approved scheme is for all phases of the mission under both nominal and degraded conditions in order that they can insure compatibility of the work they are doing. Of course, a process would have to be established for modifying these schemes as discrepancies or undesirable characteristics are uncovered. That is, some sort of change control is needed here as it is in so many other areas. Considerable work has progressed under Dick Carley's guidance to get this method of operation underway. For example, he fostered the attached TRW working paper which covers two of the major mission phases--Ascent and Rendezvous. It is their attempt at developing logic flow diagrams

describing the normal G&C functions, identifying decision points in the mission, and establishing the procedures for monitoring the G&C systems required to assess systems performance and govern subsequent action. As far as I can determine, this document reflects the work of a few TRW people with some limited input informally obtained from individuals within MSC.

3. It is evident that in order to have a truly useful working document, close coordination and cooperation of all MSC elements and our contractors is required. What I would like to do, after allowing sufficient time for those of you interested in this work to review the attached material, is to have a meeting to discuss the manner in which we will carry out the coordination of this activity. Assuming documentation of the type attached is a necessary part of this activity, as I believe it probably is, we will also review the format and technical content of at least one of the mission phases documented here. Our primary purpose would be to make sure it does the job it needs to do in the best way. Right now I think it would be reasonable to aim for a meeting on the Ascent phase in late September for that dual purpose.

4. Based on the results of that meeting, I would propose to initiate a series of meetings early in the development of the documentation such as this for the rest of the mission phases during which the opinions and inputs of everyone concerned may be discussed and included from the beginning instead of after the work is carried as far as in the attached. The other mission phases are currently broken out as follows: (a) Earth Launch - beginning early in the countdown and ending after TLI, (b) Trajectory Maneuvers including MCC, LOI, TEI, (c) Lunar Descent, and (d) Reentry. As you can see this all applies directly to the lunar landing mission, but obviously it will also have considerable bearing on how we fly the earlier development flights.

I'll be in touch with you again.

Howard W. Tindall, Jr.

Enclosure

Addressees:
(See attached list)

MISSION TECHNIQUES DEVELOPMENT UNDER TINDALL

Tindall's overall approach to mission techniques development was to first establish appropriate working groups, structure and schedule their meetings as needed and document appropriately the results from these meetings – agreements reached, actions/ open items and the rationale for decisions made. These "Data Priority" Working Groups were established for each phase of each manned mission. They were under the overall direction of Tindall, and smaller meetings were chaired by an MSC individual to work out detailed issues and provide preliminary approval of the Mission Techniques documents published.

Tindall himself conducted many working group meetings that were comprised of various representatives of MSC and MSC support contractors to coordinate their various associated activities to develop and document the agreed-upon mission techniques. From these meetings came the well-known "Tindallgrams" and the Apollo Mission Techniques documents published by TRW.

The two earlier documents created by TRW under Dr. Shea's tasking (available now at UHCL archives) were the starting point for Tindall's Data Priority effort, and on October 19, 1967, he set up the first two *data priority working groups* - Ascent/Rendezvous and Descent - to continue their development.

These were the very first of Tindall's many Mission Techniques Meetings. Review of the Ascent/Rendezvous Phase document was held October 27, 1967. This first working group consisted of only six persons as shown in Tindall's memo on page 84 (Part II).

Most of this meeting was focused on review of the eight long foldout charts contained in the Ascent/Rendezvous document (all charts were 11 inches high – 4 were 14" to 28" long, 4 were 46" to 55" long). Each presented the sequence of GNC events and decision logic to be undertaken by the three astronauts and MCC-H ground personnel for the Ascent and Rendezvous mission phases: Post-touchdown; Prelaunch Preparation; Powered Ascent into a safe orbit; safe orbit Insertion to CSI; CSI to CDH; CDH to TPI; Midcourse corrections and Braking (CSI, CDH, TPI, Midcourse corrections and Braking are the LM ΔV burns needed for rendezvous with the orbiting CSM).

Charts similar to these were used in the later documents but instead of long foldouts they were chopped into 8 ½" x 11" pages - harder to read but easier to publish and handle.

The second Mission Techniques Meeting under Tindall was held October 31, 1967 for review of the Descent Phase document. This working group consisted of the eight persons as shown here in Tindall's memo.

As with the Ascent/Rendezvous document, this meeting also focused on the charts governing lunar descent, described in four long charts (11" high by 26" to 49" long) for lunar orbit navigation updates, LM CSM Separation, Hohmann transfer and powered descent. Then on Dec 11, 1967, Mr. Tindall established the Midcourse Phase Mission Techniques Working Group, with the first meeting to be on December 12. 1967.

Tindall's intent that the descent phase mission techniques document would "become a working, configuration controlled document" is shown in the two memos sending this chapter to MSC Systems Engineering Division and MIT.

UNITED STATES GOVERNMENT

Memorandum

TO : See list below

DATE: OCT 19 1967

67-PA-T-87A

FROM : PA/Chief, Apollo Data Priority Coordination

SUBJECT: First meeting of the Ascent Phase Mission Techniques Working Group

1. This memo is in confirmation of our telephone conversation requesting your participation in a continuing working group to discuss the Ascent phase of the lunar landing mission. The overall composition of this Ascent Phase Mission Techniques Working Group is as follows:

FCD	Charlie Parker, Chairman
Flight Crew	Pete Conrad
FCSD	Paul Kramer
MPAD	Marlowe Cassetti
TRW	Dick Boudreau
ASPO	Bill Tindall

2. The first meeting is scheduled for Friday, October 27, 1967, in my office (Building 30, Room 3068) at 9:00 a.m. It's purpose is to go through the operational flow diagrams and associated analyses covered in the Ascent section of the TRW Apollo Mission Techniques document I sent you in September, and start the cyclic modification process needed to get it "right."

3. It is desirable, I think, to keep this working group as small as possible, however, if you feel it desirable to bring additional people from your organization, be my guest. Ultimately, once this document has been thoroughly reworked and substantial agreement has been obtained among the working group members on its content, it will be republished for final review with much wider participation.

Howard W. Tindall, Jr.

Addressees:
CB/P. Conrad
CF24/P. Kramer
FC/C. Parker
FM7/M. Cassetti
PA/G. Low (Info)
TRW/D. Boudreau

PA:HWTindall, Jr.:pj

UNITED STATES GOVERNMENT

Memorandum

TO : See list below

DATE: OCT 19 1967

67-PA-T-88A

FROM : PA/Chief, Apollo Data Priority Coordination

SUBJECT: First meeting of the Descent Phase Mission Techniques Working Group

1. This memo is to request your participation in the detailed development of the Apollo Mission Techniques for the Descent phase of the lunar landing mission. We are setting up a small working group of the following composition to do this job:

MPAD	Floyd Bennett, Chairman
Flight Crew	Pete Conrad
FCSD	Paul Kramer
FCD	Phil Shaffer
FSD	Tom Gibson
G&C	Myron Kayton, Ed Smith
TRW	Bob Kidd
ASPO	Bill Tindall

2. The first meeting will be held on October 31, 1967, in my office (Building 30, Room 3068) at 9:00 a.m. The purpose of this meeting will be to review and propose modifications as necessary to the attached document prepared by TRW. It is anticipated that after several such meetings the overall operational logic will have been established and documented in sufficient detail and accuracy to permit wider distribution and comment. Right now, though, it seems desirable to keep the size of this working group as small as possible. However, if you want to bring additional people from your organization, feel free to do so.

Howard W. Tindall, Jr.

Enclosure

Addressees:
CB/P. Conrad
CF24/P. Kramer
EG/M. Kayton
   E. Smith
FC/P. Shaffer
   C. Parker
FM6/F. Bennett
FM7/M. Cassetti
FS5/T. Gibson
PA/G. M. Low (Info)
TRW/B. Kidd

UNITED STATES GOVERNMENT

Memorandum

TO : System Engineering Division
Attention: PD3/R. V. Battey

DATE: OCT 19 1967
67-PA-T-90A

FROM : PA/Chief, Apollo Data Priority Coordination

SUBJECT: Descent phase to be added to your Apollo Mission Techniques document

1. TRW recently completed the attached chapter to be added into the appropriate vacancy in the Apollo Mission Techniques document I sent you recently. As was the case for the other mission phases, this section on Descent was done essentially without participation by other than TRW personnel. Accordingly, I expect it will be modified substantially before it becomes a working, configuration controlled document.

2. I am sending it to you because I thought you might find it interesting in the meantime.

Howard W. Tindall, Jr.

Enclosure

PA:HWTindall, Jr.:jj

146

NATIONAL AERONAUTICS AND SPACE ADMINISTRATION
MANNED SPACECRAFT CENTER
Houston, Texas 77058

OCT 13 1967

TO   : Massachusetts Institute of Technology
           Instrumentation Laboratory
           75 Cambridge Parkway
           Cambridge, Massachusetts. 02142
           Attention: R. Ragan

FROM   : Chief, Apollo Data Priority Coordination

SUBJECT: Descent phase to be added to your Apollo Mission Techniques
         Document

TRW recently completed the attached chapter to be added into the
appropriate vacancy in the Apollo Mission Techniques document. I
sent you recently. As was the case for the other mission phases,
this section on Descent was done essentially without participation by other than TRW personnel. Accordingly, I expect it will
be modified substantially before it becomes a working, configuration controlled document.

I am sending it to you because I thought you might find it interesting in the meantime.

Howard W. Tindall, Jr.

Enclosure

UNITED STATES GOVERNMENT

Memorandum

TO   : See list below

DATE: DEC 11 1967

67-PA-T-111A

FROM   : PA/Chief, Apollo Data Priority Coordination

SUBJECT : Establishment of the Midcourse Phase Mission Techniques Working Group

1. This memo is in confirmation of telephone conversations requesting your participation in a continuing working group to discuss the Midcourse Phase of the lunar landing mission. This mission phase includes the TLI, MCC, LOI and TEI maneuvers. The overall composition of this Midcourse Phase Mission Techniques Working Group is as follows:

FM	Ronald Berry, Chairman
CB	Mike Collins
CF	Ted Guillory
EG	Ed Smith
EG	Clarke Hackler
FC	Ed Pavelka
FC	Steve Bales
FC	Dave Massaro
TC	Gary Coen
FM	Stan Mann
FM	Tom Kyle
FM	Jerry Yencharis
PA	Bill Tindall
TH	Bob Kidd
TH	Dick Boudreau

2. The first meeting is scheduled for Tuesday, December 12, 1967, in Building 30, Room 3044, at 9:00 a.m. Its specific purpose is to establish targeting and monitoring procedures for the LOI maneuver to be included in the overall Apollo Mission Techniques document. It is desirable, I think, to keep this working group as small as possible, however, if you feel it desirable to bring additional people from your organization, be my guest.

Howard W. Tindall, Jr.

Addressees:
(See attached list)

Buy U.S. Savings Bonds Regularly on the Payroll Savings Plan

OPTIONAL FORM NO. 10
MAY 1962 EDITION
GSA GEN. REG. NO. 9

UNITED STATES GOVERNMENT

Memorandum

FROM : FM/Chief, Apollo Data Priority Coordination

SUBJECT: Data Priority meetings schedule

1. In order to get a little bit more orderly in our conduct of the Trajectory Control Data Priority business, it's evident that we must schedule meetings on a pre-established, periodic basis. Accordingly, the Ascent and Descent meetings will take place on alternate Tuesday afternoons; Rendezvous meetings will be held on Wednesdays, one mission in the morning and another in the afternoon (right now they will be missions "G" and "D", respectively) in the same week as Ascent. The so-called Midcourse Phase, name to be changed I hope, will take place on Wednesday mornings in the same week as Descent. All of these meetings will be approximately ½ day long.

2. This sounds like a rather large time absorber because it is. However, there are a number of us who feel that the only way we can get this thing done at all - to say nothing of being on time - is to make a fairly intensive effort for a short time. For example, aside from the Rendezvous business, I anticipate the rest of the mission phases should be pretty well squared away within three or four months and after that they will require much less attention.

3. One more thing regarding the Midcourse Phase——it is proposed that this panel's work be enlarged to encompass all pre-launch targeting associated with the lunar landing mission, all activities in earth orbit, the TLI maneuver, the Midcourse maneuvers, the LOI maneuver and the TEI maneuver. It will not include launch and launch abort monitoring. The reason for this is that all those activities listed, aside from TEI are aligned to the same end objective; that is, braking into a suitable lunar orbit.

Howard W. Tindall, Jr.

Addressees:
(See attached list)

Many Meetings got it done

Shortly after these initial meetings, it became evident to Mr. Tindall that a three to four month intensive effort would be required (significantly underestimated at that time - many more meetings over a longer period of time were needed). Accordingly at that time Mr. Tindall established a regular weekly meetings schedule. For example, subsequent Ascent and Descent meetings were to be on alternate Tuesday afternoons; Rendezvous would be all day Wednesday - mornings for Mission C and afternoons for Mission D, and so on.

These earlier meetings were primarily supported by the appropriate MSC in-house participants as shown here, but other meetings were called as needed to address specific issues and questions, supported by appropriate individuals from MIT, NAR, GAEC, TRW, etc.

Bill chaired the larger Data Priority meetings between astronauts, mission controllers, design engineers, contractors, and other relevant parties, adjudicating disagreements and overseeing the details of planning the mission techniques. The working groups for each mission Phase had their individual chairman, an MSC person who typically conducted smaller meetings to conduct a final review of the mission techniques documents prepared by TRW. Tindall himself explains these meetings in his "Techniques of Controlling the Trajectory" paper:

> "Of course, the way we got this job done was with meetings - big meetings, little meetings, hundreds of meetings! The thing we always tried to do in these meetings was to encourage everyone, no matter how shy, to speak out, hopefully (but not always) without being subjected to ridicule. We wanted to make sure we had not overlooked any legitimate input.
>
> One thing we found to be very effective-and which we almost always did - was to make decisions on how to do the job, even if the data available were incomplete or conflicting, or if there was substantial disagreement among the participants. This even included making educated guesses at the performance abort limits and stating that they were the values we would use unless someone came in with something better. However, do not get me wrong. These limits are a very serious business.

They literally define the point at which a mission will be aborted. You can imagine how emotional our meetings frequently were!

Although the decisions reached often displeased someone, the fact that a decision had been made was invaluable. Since this effort was officially sanctioned, the decisions served to unify all subsequent work. They often also pointed up the need for more work. And for those dissatisfied with the decisions, at least they presented a firm target which they could attack through recognized channels to higher authority.

In short, the primary purpose of these meetings was to make decisions, and we never hesitated! These early decisions provided a point of departure; and by the time the flight took place, our numbers were firm, checked, and double-checked. By then, we knew they were right!

The meetings were regularly attended by experts involved in all facets of trajectory control systems, computer, and operations people, including the crew. Our discussions not only resulted in agreement among everyone as to how we planned to do the job and why, but also inevitably educated everyone as to precisely how the systems themselves work, down to the last detail. A characteristic of Apollo you could not help noting was just how great the lack of detailed and absolute comprehension are on a program of this magnitude. There is a basic communication problem for which I can offer no acceptable solution. To do our job, we needed a level of detailed understanding of the functioning of systems and software far greater than was generally available. Through our meetings, however, we forced this understanding. It was not easy, but we got it sorted out eventually – together."

Tindall's management style as described in Woods' book *How Apollo Flew to the Moon*.

"There were two sides to his style. The first was the manner in which he handled large meetings that involved engineers, programmers, mathematicians, crews or whoever in order to get this diverse mass of people to reach a decision – 'knocking people's heads together,' as one engineer described it.

Tindall would control the debates in terms of giving people the opportunity to talk, and then mix and match and make the trades. Then he would make a decision and say, 'I'm gonna recommend this to management. Anybody have any really strong objections?' And the guy who lost the debate may say 'Yeah it won't work!' And Tindall would say, 'OK, fine. We'll go this way if it won't work, we'll come back and re-address it, but we'll make a decision today.'

They were good debates and anybody could stand up and debate the issue. But he kept it moving. He didn't get bogged down because he himself was a brilliant engineer. I think Tindall was a real key to the success of Apollo because of how he brought people together and had them communicate in very complex issues. He was very good at it. He'd have them explain it, and in front of all their peers.

The second side to Tindall's ability was in the extraordinary memos he wrote, now fondly called **Tindallgrams** - they revealed a unique chatty, easy to understand style that historians thought was quite remarkable. For example, a memo that discussed the possible reasons for Apollo 11's overshoot had as its subject line, 'Vent bent, 'descent lament!'"

TRW, as staff to Tindall, participated in these meetings as required to develop the logic flow diagrams and write various other sections of the Mission Techniques documents. We coordinated inputs from the relevant MSC organizations and produced draft documents for review at smaller working group meetings. Next after a final formal review by ASPO, E&DD, FCOD, FOD, GAEC, MIT, NR, and TRW, we prepared the final masters, obtained Mr. Tindall's cover letter signature, accomplished their printing and then delivered the required number of copies to NASA/MSC for their distribution. We also maintained a list of open items/actions from the various meetings and provided these to Mr. Tindall for his use in follow-on meetings.

Logic Flow Charts – document real-time decisions

All Mission Techniques documents covering a mission phase contain logical flow diagrams covering many pages. They describe both the nominal/real-time sequence of GNC events (State Vector updates, IMU drift checks and alignments, ΔV uploads, monitoring and checks for onboard GNC health, etc.) and the decision logic for flight crew and ground controllers (one flow line for each), and the procedures to be followed if certain limits shown on the charts are exceeded.

The decision logic and limits shown are not meant to be followed exactly, step by step, but rather provide guidance during simulated and real-time flight operations as well as overall direction for flight and ground crew procedures development.

One of the larger (and longer) Mission Techniques meetings was held February 12, 1968, covering the D Mission. There were 5 astronauts at this meeting (Aldrin, McDivitt, Schweickart, Stafford and Young), as well as 18 MSC engineers, Tindall, myself and one other TRW engineer. At this meeting TRW handed out for review the flow diagrams for the "D" Mission Rendezvous; however, the meeting became long and spirited, so they were never reviewed. Tindall's minutes from the meeting:

> 4. TRW distributed copies of their first cut at the mission techniques logical flow for the rendezvous exercise up to and including the so-called insertion maneuver made by the LM to break out of the initial football rendezvous trajectory. These flow charts will probably be reviewed in detail at the next meting.

The example flow chart shown here was taken from the Lunar Surface Phase mission techniques document, which covers GNC activities from LM landing to liftoff. This chart shows only a portion of the ground and spacecraft activities needed prior to liftoff from the moon. Note that it used separate lines for MCC-H and the three flight crew members, typical of most charts.

The flow diagrams were published in a top down format as shown here for ease in publication and handling. The first mission techniques documents used long foldout charts (11 ½ " in height by anywhere from 14" to 55" in length – convenient for group review with time running from left to right as familiar, but printing and folding of the long sheets would be quite costly for the hundreds of copies needed.

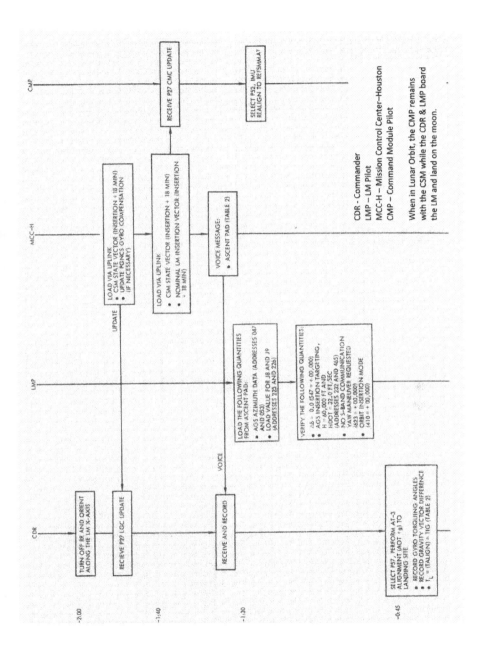

CDR - Commander
LMP – LM Pilot
MCC-H – Mission Control Center—Houston
CMP – Command Module Pilot

When in Lunar Orbit, the CMP remains
with the CSM while the CDR & LMP board
the LM and land on the moon.

Tindallgrams

After each of his many meetings, Tindall wrote minutes that very clearly described what was decided and why, what the open issues were and what further actions were to be taken. His memos were noted for their clarity and folksy writing style.

Astronaut Ken Mattingly's comments on the Tindallgrams - JSC oral interview[13]

"These meetings would go on sometimes from eight in the morning until eight in the evening, whatever it took. Room filled with people - Bill was after answers."

"And the next day you would get this two maybe three-page memorandum from Bill written in a folksy style, saying, 'You know, we had this meeting yesterday. We were trying to ask this. If I heard you right, here's what I think you said and here's what I think we should do.' And he could summarize these complex technical and human issues and put it down in a readable style. I mean, people waited for the next Tindallgram. That was like waiting for the newspaper in the morning. They looked forward to it."

---

[13] This is from an extract from Ken Mattingly's interview on November 6, 2001, for the NASA JSC Oral History Project. Ken Mattingly was Command Module Pilot for Apollo 16.

The Apollo Spacecraft Software CCB

In addition to his assignment as Chief, Apollo Data Priority Coordination, George Low also made Tindall responsible for MIT's onboard software development. Wearing this other hat as Deputy Chief of MPAD, he participated in Apollo Spacecraft Software CCB meetings and wrote memos to document key aspects of these meetings.

To ensure TRW's mission techniques work kept up with changes to the onboard software, I was on Tindall's distribution list for any Spacecraft Software CCB results applicable to the mission techniques work. As an example, the following memo is indicative of his depth of understanding and sometimes humorous writing style – note the third paragraph on the second page - one of my favorites:

> "MIT was requested to determine the impact of changing the descent program such that it would be possible for the crew to command all four RCS jets in the minus X direction immediately upon touchdown in order to smoosh the LM into the lunar surface and keep it from turning over while the DPS belches to a stop. Ain't that the damnest thing you ever heard?"

OPTIONAL FORM NO. 10
MAY 1962 EDITION
GSA FPMR (41 CFR) 101-11.6

UNITED STATES GOVERNMENT

Memorandum

TO : See list below

DATE: September 23, 1968

68-FM-T-201

FROM : FM/Deputy Chief

SUBJECT: Results of September 17 Apollo Spacecraft Software Configuration Control Board (ASSCCB) meeting

The first three hours of this marathon meeting were devoted to implementation of the descent program in LUMINARY. The currently approved plan is to implement the one-phase descent scheme proposed by Floyd Bennett and his merry crew. However, MIT has been directed to implement it in such a way that it would be possible to fly the old two-phase technique - if desired. Almost all effort is to be devoted to the one-phase technique with only one day's worth of testing included for the two-phase - and no design improvements are to be developed or included in the two-phase. What this really means is that at the cost of one day's worth of testing we have provided some cheap insurance for being able to change back later if we have to. If the decision were made to use the two-phase, a considerable amount of additional testing would be required and at that time, program deficiencies might be uncovered revealing that that capability does not really exist.

Several things that interested me about the new one-phase are:

    1. The decision of which way to go - one or two-phase is made pre-flight and an option flag is set in erasible memory before launch.

    2. The much smoother attitude time history of the one-phase scheme may very well permit the DPS trim gimbal to do all the steering, substantially reducing RCS usage.

    3. MIT is providing a crew option via the DSKY for manually changing from P63 to P64 in the event they want to do that earlier than the automatic switch.

    4. High-gate is now being defined as the time at which the landing radar position is changed.

MPAD has submitted a Program Change Request (PCR 249) to eliminate a lock-out of the landing radar data above 35,000 feet (estimated altitude). This was a two part change since it is necessary to fix a program to allow the data to be read and also necessary to change the weighting function such that data above 35,000 feet is not given a zero influence

on the state vector. Since the proposed change was estimated to cost
three days schedule impact, Floyd Bennett was requested to rewrite his
PCR to simplify the requirement while achieving the same end results.
Essentially, it amounted to replacing the 35,000 foot boundary with a
50,000 foot boundary. In addition, it is necessary that I verify that
the rendezvous radar powered flight designate routine (R29) can be
eliminated as a requirement and thus be made uncallable from the descent
programs. Subsequent to the meeting I did that and have informed FSD.

Guidance and Control Division brought in two PCR's (Nos. 224 and 248)
which influence the processing of the landing radar data. One changed
the reasonability tests and the other provided a delay in utilizing
landing radar data for four seconds after the LGC receives a "data
good" discrete because it takes that long for the landing radar output
to converge on the true value after lock-on. Both were approved at a
cost of one day each.

MIT was requested to determine the impact of changing the descent program
such that it would be possible for the crew to command all four RCS jets
in the minus X direction immediately upon touchdown in order to smoosh
the LM into the lunar surface and keep it from turning over while the
DPS belches to a stop. Ain't that the damnest thing you ever heard?

Flight Crew Support Division presented a proposal to modify COLOSSUS II
to permit the crew to manually steer the TLI burn in the event of a
SIVB IU failure. No action will be taken on this until the technique
is approved by Mr. Low's CCB.

A really ancient PCR, No. 132, submitted by the crew to provide a VHF
ranging data good discrete light, was finally disapproved since the
spacecraft will not be modified to provide the additional DSKY lights
which would have been used for this.

Tom Gibson presented their proposal, which was approved, for the follow-
on spacecraft programs. A so-called COLOSSUS I Mod A will be prepared,
which is basically the COLOSSUS I program with all known anomalies
corrected plus the following three simple program improvements:

1. IMU pulse torquing

2. Backup integration

3. An improvement on the mark incorporation.

It is planned that a tape release of this program will occur on December
1, at which time mission operations testing (Level 6) can be started
along with rope manufacture. This program will be used for the D mission.

158

A COLOSSUS II program is also now being developed which starts from the COLOSSUS I Mod A baseline to which CSI/CDH will be added. I suppose it will also include anomalies uncovered too late for the Mod A version. MIT's estimate of tape release for this program is February 1, 1969. It is felt that this program can probably be made ready for Spacecraft 106 - that is, the flight after D, whatever that is. VHF ranging, incidentally, should also be available on spacecraft 106.

Howard W. Tindall, Jr.

Addressees:
FM/J. P. Mayer
   C. R. Huss
   D. H. Owen
FM13/R. P. Parten
   J. R. Gurley
   E. D. Murrah
   M. Collins
FM4/P. T. Pixley
   R. T. Savely
FM5/R. E. Ernull
FM5/H. D. Beck
FM6/R. R. Regelbrugge
   K. A. Young
FM7/S. P. Mann
   R. O. Nobles
FM/Branch Chiefs
TRW/Houston/R. J. Boudreau
MIT/IL/M. W. Johnston

FM:HWTindall, Jr.:js

A Controlled Baseline Provided Guidance for Many

As an officially approved and controlled baseline, the Mission Techniques documents were recognized by top-level NASA management as highly important to the effective coordination of the efforts of the many involved organizations. TRW had emphasized from the beginning, starting with the work done under Dr. Shea, the critical need for a formally controlled and recognized baseline of GNC events and data priority decisions.

Responding to this, eight months after his start Tindall requested program office clarification on the status of his efforts – specifically on the priority of mission techniques documentation vis-à-vis other official Apollo program documentation.

Following a meeting of senior MSC managers to address this issue, ASPO manager George Low issued the memo shown next stating that mission techniques documents were to be considered as controlled documentation, and that they had precedence over Mission Rules, the AOH (Apollo Operations Handbook), flight crew procedures and flight plans – for all Apollo manned missions. This further ensured compatibility of the work being performed by the many organizations and individuals associated with the Apollo program.

Tindall's "Techniques of Controlling the Trajectory" paper emphasizes these wide-ranging impacts:

> "Since trajectory-control activities make up such a large part of every Apollo mission, the manner of conducting them has an impact on almost every other facet of the mission, even on aspects that seem remote, such as electrical-power management, thermal control, and consumable budgeting. As a result, it is necessary to consider all of these things in the development of the overall trajectory-control procedures, which we call mission techniques. Also, *it is important that almost everyone in the Apollo world knows how we intend to do these things, since we often impact their plans.*"

The Apollo World

To illustrate the size of this "Apollo world, shown here is my TRW transmittal letter to Tindall for the "final" (pre-Apollo 11) version of the Lunar Descent Mission Techniques that I had saved from my Apollo days.

For this particular delivery we printed 375 copies of each document which included Tindall's cover letter and his distribution list on top, then delivered these to NASA/MSC for their distribution. A very large world indeed as shown by the many MSC and contractor individuals on this list.

I also have an internal distribution for this document which shows 80 TRW individuals at Houston Operations, LA (Redondo Beach, Space Park) and KSC.

OPTIONAL FORM NO. 10
MAY 1962 EDITION
GSA GEN. REG. NO. 27
5010-107

UNITED STATES GOVERNMENT

Memorandum

TO : PA/Chief, Apollo Data Priority Coordination

FROM : PA/Manager, Apollo Spacecraft Program

DATE: May 2, 1968
In reply refer to:
PA-8-5-6

SUBJECT: Status of data priority documentation

On May 1, 1968, a meeting was held in my office to discuss control procedures and precedence of your data priority documentation. Participants in this meeting were Kraft, Slayton, North, you, and I. The purpose of this memorandum is to document the decisions reached in the meeting.

It was agreed that:

a. The data priority documentation will be considered as controlled documentation and will have precedence over Mission Rules, the AOH, flight crew procedures, and flight plans.

b. You will participate as a member on the Crew Procedures Control Board. If in the course of a Crew Procedures Control Board meeting a crew procedure in violation of the data priority documentation is proposed, one of the two following courses of action is open:

(1) If you concur in the proposed change, the change may be approved by the Crew Procedures Control Board and you will update the data priority documentation accordingly.

(2) If you do not concur in the proposed change, the Crew Procedures Control Board will not approve the change but will, instead, bring it to the Apollo Spacecraft Program Office Configuration Control Board.

c. At some point in time prior to each mission, the data priority documentation will no longer be updated and other documents will become controlling. It was generally agreed that there should be at least a 2 months' overlap where both the data priority documentation and the other documentation are valid, however. An action item was assigned to you to determine whether all items contained in the data priority documentation are indeed reflected in other documents and when control of the data priority documentation should be discontinued.

George M. Low

cc:
CA/D. K. Slayton
CF/W. J. North
FA/C. C. Kraft, Jr.
PA/G. W. S. Abbey

PA:GMLow:jsw 5-2-68

TRW No. 69:7252.4-116
Re: Task ASPO 46C

23 June 1969

National Aeronautics and Space Administration
Manned Spacecraft Center
Houston, Texas 77058

Attention: Mr. H. W. Tindall, Jr. (FM)

Subject: Task ASPO 46C Submittal

Gentlemen:

Forwarded to you under separate cover are 375 copies of the "Apollo
Mission Techniques, Mission G, Lunar Descent, Revision A - Techniques
Description," (MSC Internal Note No. S-PA-8N-021A).

This submittal presents the officially approved Mission Techniques.
Several open items are noted in the text. The attached table presents
the description of the open items and the page number for reference
purposes.

Very truly yours,

K. J. Boudreau, Task Manager
Task ASPO 46C

V. R. Widerquist
Assistant Project Manager (Acting)
Apollo Spacecraft Program Support

WJK:RJB:lp

Distribution:

Mr. Ralph Albon (PP7)
Mr. R. B. Battey (PD)
Apollo Document Distribution (BF66)
Contracting Officer, Flight Operations
   Procurement Section (BG671)
Technology Utilization Office (BF34)

DISTRIBUTION LIST

AA/R. R. Gilruth
AB/G. S. Trimble
CA/D. K. Slayton
CB/Astronaut Office (48)
CF/W. J. North
CF13/D. F. Grimm
CF212/C. Jacobsen
CF212/W. Haufler
CF212/W. Hinton
CF212/J. L. Blanco
CF2/J. Bilodeau
CF22/D. L. Bentley
CF22/R. L. Hahne
CF22/M. C. Gremillion
CF22/W. B. Leverich
CF22/C. C. Thomas
CF24/P. Kramer
CF24/J. Rippey
CF24/M. C. Contella
CF24/D. W. Lewis
CF24/D. K. Mosel
CF32/J. J. Van Bockel
CF32/M. F. Griffin
CF33/M. Brown
CF33/C. Nelson
CF34/T. W. Holloway
CF34/L. J. Riche
CF34/J. V. Rivers
CF34/T. A. Guillory
CF34/E. B. Pippert
CF34/C. L. Stough
EA/M. A. Faget
EA2/J. B. Lee
EA4/J. Chamberlin
EA5/P. M. Deans
EB/P. Vavra
EE/L. Packham
EE/R. Sawyer
EE13/M. J. Kingsley
EE13/R. G. Irvin
EE3/R. L. Chicoine
EE6/G. B. Gibson
EE6/R. G. Fenner
EE6/J. R. McCown
EG/R. J. Chilton
EG13/W. J. Klinar
EG2/C. T. Hackler
EG23/K. J. Cox
EG23/E. E. Smith
EG25/T. V. Chambers
EP2/W. R. Hammock
EA2/R. A. Gardiner
AP3/R. T. White
FC/G. D. Griffin

EG27/W. R. Warrenburg (2)
EG27/H. E. Smith
EG41/J. Hanaway
EG42/B. Reina
EG43/A. R. Turley
EG44/C. W. Frasier
EG/MIT/T. Lawton
KA/R. F. Thompson
PA/G. M. Low
PA/C. H. Bolender
PA/K. S. Kleinknecht
PA2/M. S. Henderson
PB/A. Hobokan
PC/W. H. Gray
PD/O. E. Maynard
PD/C. D. Perrine
PD/R. V. Battey
PD12/J. G. Zarcaro
PD12/R. J. Ward
PD12/R. W. Kubicki
PD12/J. Sevier
PD4/A. Cohen
PD6/H. Byington
PD7/W. R. Morrison
HA/J. P. Loftus
PE/D. T. Lockard
TH3/J. E. Dornbach
TJ/J. H. Sasser
CO7/J. Nowakowski
FA/C. C. Kraft, Jr.
FA/S. A. Sjoberg
FA/C. C. Critzos
FA/R. J. Rose
FA4/C. R. Hicks
FC/E. F. Kranz
FC/M. P. Frank
FC/G. S. Lunney
FC/C. E. Charlesworth
FC/M. Windler
FC/J. Roach
FC2/C. S. Harlan
FC2/H. M. Draughon
FC2/J. H. Temple
FC25/C. R. Lewis
FC27/W. E. Platt
FC27/H. Johnson
FC27/L. Pavlow
FC3/A. D. Aldrich
FC3/N. B. Hutchinson
FC35/B. N. Willoughby
FC35/G. E. Coen
CF3/C. H. Woodling
AP3/M. E. Reim (2)
FC35/R. Fruend

FC44/R. L. Carlton
FC44/J. B. Craven
FC44/J. C. Watros
FC5/J. C. Bostick
FC5/P. C. Shaffer
FC54/J. S. Llewellyn
FC54/C. F. Deiterich
FC54/J. E. I'Anson
FC55/E. L. Pavelka
FC55/G. C. Guthrie
FC35/C. T. Essmeier
FC55/H. D. Reed
FC55/M. G. Kennedy
FC55/J. H. Greene
FC56/C. B. Parker (4)
FC35/W. J. Strahle
FC35/G. P. Walsh
FC4/J. E. Hannigan
FC6/C. B. Shelley (4)
FL/J. B. Hammack
FL2/R. L. Brown (2)
FL6/R. W. Blakley
FS/L. C. Dunseith
FS5/J. C. Stokes
FS5/T. F. Gibson
FS5/J. L. Hall
FS5/C. R. Parker
FS5/C. A. Lander
FS5/L. J. Dungan
FS5/J. E. Williams
FS5/T. G. Price
FM/J. P. Mayer
FM/H. W. Tindall (15)
FM/C. R. Huss
FM/D. H. Owen
FM/R. P. Parten
FM13/W. J. Bennett
FM13/F. J. Suler
FM13/J. R. Gurley
FM13/C. W. Pace
FM13/M. A. Collins
FM13/K. H. Henley
FM13/J. Michos (2)
FM13/W. Lacey
FM13/B. F. McCreary
FM2/F. V. Bennett
FM2/C. A. Graves
FM2/J. C. Harpold
FM2/R. D. Nelson
FM3/C. Allen
FM3/C. T. Hyle
FS5/G. R. Sabionski
FS5/M. Conway
FM6/H. L. Conway
FM6/R. S. Merrian
CF22/T. H. Kiser
FM4/E. R. Schiesser

FM4/H. G. deVezin
FM4/A. D. Wylie
FM4/J. C. McPherson
FM4/P. T. Pixley (2)
FM4/R. T. Savely
FM4/W. R. Wollenhaupt
FM5/R. L. Berry
FM5/R. E. Ernull
FM52/J. R. Elk
FM52/T. J. Linbeck
FM52/K. T. Zeiler
FM52/Q. S. Holmes
FM54/J. D. Yencharis (4)
FM55/H. D. Beck
FM55/R. D. Duncan
FM55/G. W. Ricks
FM6/E. C. Lineberry
FM62/R. W. Becker
FM62/J. A. Bell
FM64/K. A. Young
FM64/J. D. Alexander
FM64/A. H. Benney, Jr.
FM64/D. D. DeAtkine
FM64/R. H. Moore
FM7/M. D. Cassetti
FM7/S. P. Mann
FM7/R. O. Nobles
FM7/D. A. Nelson
FM8/J. Funk
BM2/Technical Library (2)
BOEING (Houston)

HH-02/R. B. McMurdo (2)
HA-58/R. L. Allen
HM-25/H. E. Dornak
HM-25/D. W. Hackbart
HM-08/D. Heuer
BELLCOMM, INC.,
955 L'Enfant Plaza North, S. W.
Washington, D. C. 20024

R. V. Sperry
G. Heffron
D. Corey
A. Merritt (MAS)

GRUMMAN AIRCRAFT ENGINEERING CORPORATION
Bethpage, Long Island
New York   11714

W. Obert-Thorn
R. Schendwolf (3)
R. Mangulis
R. Pratt (15)
B. O'Neal
Consulting Pilot's Office

MASSACHUSETTS INSTITUTE OF TECHNOLOGY
Instrumentation Laboratory
75 Cambridge Parkway
Cambridge, Mass.   02142

Ralph R. Ragan (25)
Malcolm W. Johnston

NORTH AMERICAN ROCKWELL CORPORATION
             SPACE DIVISION
12214 Lakewood Boulevard
Downey, California   90241

M. Vucelic, FB84   D.W. Patterson, AC50
R. Zermuehlen, FB59
J. E. Roberts, FB59
B. C. Johnson (4), AB46
W. H. Markarin, FB55
E. Dimitruk, BB49
J. E. McIntyre, BB48
M. B. Chase, AB33
NATIONAL AERONAUTICS AND SPACE ADMINISTRATION
Goddard Space Flight Center
Greenbelt, Maryland   20771

F. O. Vonbun, 550

NASA HEADQUARTERS
Washington, D. C.   20546

R. B. Sheridan, MAO   Robert Sherrod, XS
R. O. Aller (2), MAOP
Colonel T. McMullen, MAO

John F. Kennedy Space Center NASA
Kennedy Space Center, Florida   32899

R. D. McCafferty, CFK ; C. Floyd, CFK; M. Walters, CFK
P. Baker, CFK
Chet Lee, MSOB Building

MITRE Corporation
Beta Building
16915 El Camino Real
Houston, Texas   77058

W. P. Kincy

A C ELECTRONICS
Oak Creek Wisconsin,   53172

E. Radsack

IBM (Houston)

G. Carlow, D79

GENERAL ELECTRIC CO.
P. O. Box 2500
Daytona Beach, Fla.

R. L. Coleman, Room A-243

GRUMMAN AIRCRAFT ENGINEERING CORPORATION
1740 NASA Blvd.
Houston, Texas 77058

Mr. G. Kingsley

OPTIONAL FORM NO. 10
MAY 1962 EDITION
GSA FPMR (41 CFR) 101-11.6

UNITED STATES GOVERNMENT

Memorandum   NASA Manned Spacecraft Center

TO   : See list attached

FROM : PA/Chief, Apollo Data Priority Coordination

SUBJECT: Mission G Lunar Descent Document

DATE: June 23, 1969
69-PA-T-96A

Attached is the Final version of the long-awaited Apollo Mission
Techniques Document for Mission G Lunar Descent.

Howard W. Tindall, Jr.

Enclosure

PA:HWT:js

Buy U.S. Savings Bonds Regularly on the Payroll Savings Plan

NATIONAL AERONAUTICS AND SPACE ADMINISTRATION

MSC INTERNAL NOTE NO. S-PA-8N-021 A

JUNE 23, 1969

APOLLO MISSION TECHNIQUES
MISSION G LUNAR DESCENT
REVISION A

TECHNIQUES DESCRIPTION

APOLLO SPACECRAFT PROGRAM OFFICE
MANNED SPACECRAFT CENTER
HOUSTON,TEXAS

APOLLO MISSION TECHNIQUES DOCUMENTS

Mission techniques documents were developed for each of the eleven manned Apollo missions; four missions (Apollo 7, 8, 9 and 10) before the first lunar landing mission (Apollo 11) and six missions after that (Apollo 12, 13, 14, 15, 16 and 17).

Documents Covered All Manned "Mission Types"

The title page of each mission techniques document first identifies the applicable Mission Type, then the applicable mission phase.

For the Apollo program, there were nine mission types identified as follows:

A – Unmanned Saturn V and CSM development flights

B – Unmanned Lunar Module development flights

C – Manned CSM in Low Earth Orbit - S-IVB rendezvous - Apollo 7

C-Prime – Manned lunar orbit flight with CSM only – Apollo 8

D – Manned CSM and LM operations in low Earth orbit – Apollo 9

E – Manned CSM and LM simulated lunar mission in a highly elliptical Earth orbit – never flown

F – Manned CSM and LM operations in lunar orbit, a "dress rehearsal" for the first landing – Apollo 10

G – First manned lunar landing – Apollo 11

H – Precision landing, 2-day stays, two "moonwalks" – Apollo 12, 13, 14

J – 3-day lunar stays, use of a Lunar Roving Vehicle – Apollo 15,16, 17

Apollo 8 was to be flown as a D mission type, but the LM was not ready in time for the December 1968 launch, so a C Prime "CSM-only in lunar orbit" mission was defined and flown for Apollo 8. The first LM flight in earth orbit (the D mission) was flown later in March 1969 - Apollo 9.

<u>Documents that governed the first moon landing</u>

The following Apollo Mission Techniques documents were in effect at the time of the first lunar landing on July 20, 1969.

<u>DOCUMENT</u>	<u>REPORT</u>	<u>DATE</u>
1. Launch Phase Aborts	SP-PA-9T-048	3/31/1969
2. Earth Parking Orbit and Translunar Injection	SP-PA-9T-144	7/14/1969
3. Translunar Midcourse Corrections & Lunar Orbit Insertion	SP-PA-9T-41	2/17/1969
3a. Change Pages for SP-PA-9T-41	69-PA-T-90A	6/10/1969
4. Lunar Orbit Activities	SP-PA-9T-135	6/30/1969
5. Lunar Descent	SP-PA-8N-021A	6/23/1969
6. Descent Abort and Subsequent Rendezvous	SP-PA-9T-137	6/30/1969
7. Lunar Surface Phase	SP-PA-9T-050	5/12/1969
8. Lunar Powered Ascent	SP-PA-9T-128	6/25/1969
9. Transearth Injection, Midcourse Corrections, and Entry	SP-PA-8T-028	10/28/1968
10. Tracking Data Select Controllers Procedures	SP-PA-9T-049	4/9/1969
11. Contingency Procedures	SP-PA-9T-043	3/24/1969
12. Manual Ascent	SP-PA-9T-055	7/17/1969

These 12 documents exist in original paper format at the University of Houston Clear Lake (UHCL) archives. There are about 50 total Mission Techniques Documents in the UHCL Archives covering the various mission types. Many have been scanned into PDF format for easy availability to space history researchers.

Thanks to Jean Grant UHCL Archivist; she provided me all the above documents in PDF format to supplement the 7 original (paper) documents saved from my Houston days.

A Typical Mission Techniques Document

Here is a typical cover and signature page that is typical of the 12 documents produced for the G Mission.

Following the cover and signature page is the standard forward page:

FORWARD

This document presents the officially approved guidance and control sequence of events, the data flow, and the real-time decision logic for aborts and subsequent rendezvous initiated between descent orbit insertion (DOI) plus 10 minutes and the second lunar stay/no stay lift-off time during the G mission.

The purpose of this document is to insure compatibility of all related MSC and supporting contractor activities. For each mission phase, a Data Priority Working Group has been established under the direction of Chief, Apollo Data Priority Coordination, ASPO. These groups,' which are comprised of representatives of MSC and support contractors, hold frequent meetings to coordinate their various associated activities and develop agreed-upon mission techniques. TRW assists in the development and documentation of techniques for ASPO. After formal review by ASPO, E&DD, FCOD, FOD, GAEC, MDC, MIT, NR, and TRW, a document such as this one is issued.

Following the forward section, but preceding the Introduction, would be the Table of Contents and as appropriate lists of tables, illustrations, nomenclature, etc., - not shown here.

The Introduction for this document is:

1. INTRODUCTION

The guidance and navigation sequence of events, the flow of data, and the real-time decisions which must be made for aborts prior to or during the powered-descent and early lunar surface phases of the lunar landing mission are presented in this document. The recommended procedures are presented in flow charts (Section 4), which are preceded by a description of the techniques. These charts document the standard, approved procedures to be used in the event of a primary guidance, navigation, and control system (PGNCS) failure, a descent propulsion system (DPS) failure, or some other critical system failure.

Routine procedures are included in the flow charts, as required, to verify the continued operation of unfailed PGNCS or DPS during an abort so that the proper alternative action will be taken in the event of subsequent malfunction. The flow charts contain a block near each decision point which indicates the responsibility for determining the decision point data limit and the date of its availability.

TRAJECTORIES FOR THE EARLY LUNAR LANDING FLIGHTS

The orbital schemes presented in this section are representative of those used for the two early lunar landing missions, Apollo 11 and Apollo 12. Later missions wherein the CSM more actively participated in Lunar Descent and Rendezvous are not included.

Earth parking orbit and Translunar Injection

The Saturn V launch vehicle placed the Apollo CSM/LM spacecraft assembly into a nominal 115 n. mile circular earth orbit. Then after about one-and-a-half revolutions the Saturn V S-IVB stage ignited for a second burn, the translunar injection (TLI) burn of almost 6 minutes, to place the Apollo spacecraft into a translunar orbit.

Free Return Trajectories

The translunar injection burn for the early lunar missions placed the Apollo CSM/LM spacecraft onto a "free-return" trajectory defined as follows:

- The spacecraft would pass behind the moon as needed to accomplish the Lunar Orbit Insertion (LOI) burn into lunar orbit

- If the LOI burn was not done (due to a system failure during translunar coasting), the spacecraft would be on a return trajectory suitable for reentry. A truly perfect free-return trajectory would need no subsequent propulsive burns, however, to correct for trajectory perturbations, small, nominally zero mid-course ΔV maneuvers are planned at specific times and executed as needed.

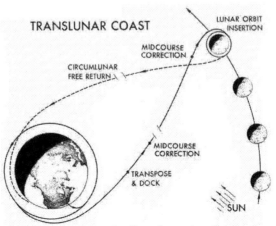

Free return trajectories were used for the first three Apollo lunar missions:

- Apollo 8 wherein the CSM-only spacecraft went into orbit around the moon.

- Apollo 10 the dress rehearsal flight which did all maneuvers of a lunar landing except the final 12 minutes of DPS thrusting to actually land the LM on the moon.

- Apollo 11 the first lunar landing mission.

Fortunately these three Apollo missions did not have to take advantage of the free return as all was well upon arrival at the moon and therefore they inserted into lunar orbit.

Hybrid Trajectories

Starting with Apollo 12, a hybrid trajectory was used to save spacecraft fuel in reaching a desired landing site on the moon – with optimum landing site lighting at the time of Lunar Descent and also permitting the Goldstone, California tracking station to monitor the LM descent and landing. For these missions, the TLI burns were targeted to place the spacecraft into a highly elliptical Earth orbit that fell short of the moon, but was effectively on a free return path to the atmospheric entry corridor.

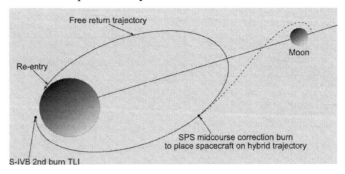

Then a mid-course maneuver was performed about 30 hours into the mission to change to a translunar trajectory that was not a free return but targeted to optimize reaching the desired landing site. This trajectory would meet flight crew and ground visibility needs for example, yet also retained the safety characteristics of initially being on a free return path and only departed from free return once the lunar module was docked with the command module (providing back-up maneuver capability) and after all systems had been checked out.

Apollo 13

Apollo 13 was the only Apollo mission to actually go around the Moon in a free-return trajectory. The oxygen tank explosion had occurred well into the mission, about a day after the CSM/LM spacecraft had transferred from the highly elliptical free-return orbit to the hybrid non-free return trajectory, and a decision was then made to transfer the crippled spacecraft back onto a free-return trajectory. Due to concerns regarding availability of electrical power and possible SM damage from the explosion, the SPS could not be used, so about 5-6 hours after the explosion this burn was accomplished using the LM DPS.

After this burn, Flight Controllers had about 18 hours to evaluate several trajectory options and determine the best option to proceed. This resulted in scheduling a second DPS burn to take place two hours after pericynthion - after the spacecraft had emerged on the free-return trajectory from behind the moon.

This burn resulted in a trajectory that shortened the return to earth time by 10 hours and moved the free-return landing spot from the Indian Ocean to the Pacific Ocean.

The LM DPS was also used a third time for a later course correction burn prior to reentry into the earth's atmosphere.

As expected, Bill Tindall played a key role in the real-time decisions needed to re-plan a safe return to earth trajectory using available propulsion and other resources.

APOLLO 13 RE-PLANNING – BILL TINDALL IN THE MIDDLE WITH HEADPHONES

Translunar MCC's & Lunar Orbit Insertion

The translunar coast time, from TLI burn cutoff until insertion into Lunar Orbit is approximately three days. During this time, transposition and docking is first; maneuvering the CSM to dock with the LM and extract it from the Saturn launch vehicle. Then checkout of the LM and CSM systems including IMU alignments, state vector updates, etc. are accomplished. Midcourse Correction burns if needed are done with the SPS or RCS during this three day period.

MCC-H is primary for spacecraft navigation and burn targeting during the coast phase. They determine any midcourse corrections that may be needed and do targeting for the LOI burns. The Apollo Command Module Computer (CMC) External Delta V (ΔV) mode was used to execute these.

When approaching the moon, the Apollo spacecraft is on a hyperbolic trajectory relative to the moon with a pericynthion of about 60 – 70 n. miles. A significant SPS retrograde delta-V, about 2,900 fps, was therefore required for the first LOI burn (LOI-1) resulting in a lunar orbit of about 70 x 190 n. miles.

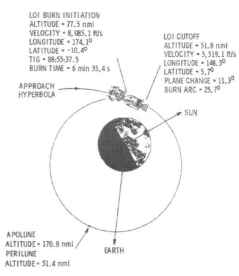

Both the LOI-1 and LOI-2 SPS burns are done when the Apollo spacecraft is behind the moon, out of sight and communications with MCC-H.

Orbits & Velocities shown for Apollo 17

Two burns were used to minimize the possibility of an over-burn resulting in an unsafe orbit. The LOI-2 burn is done 4 hours after LOI-1 (after 2 revolutions – a lunar-orbiting spacecraft period is 2 hours)

For LOI-1, the GNCS would have been shut down manually if the expected burn time (about 6 minutes) was exceeded by 9 seconds. This limit makes it extremely probable that the GNCS would automatically shut down the SPS if it was going to, and also assures a safe perilune if it didn't.

Similarly, the LOI-2 SPS burn (about 17 seconds, ΔV about 135 fps) resulting in a 60 to 70 n .mile circular orbit would have been manually shut down if the burn time had exceeded the expected time by one second.

It should be noted here that the Entry Monitor System (EMS) in the Command Module had a longitudinal accelerometer driving a digital display of accumulated ΔV along the CSM longitudinal or X-axis. After being set to zero or another initial ΔV value just before an SPS (or RCS) burn, it could be used by the flight crew as another source of information for crew-commanded SPS shutdown if needed to assure a safe lunar orbit. It was also used for ΔV monitoring and control of small RCS burns, an example of equipment designed for one purpose being used for another.

For missions after Apollo 12, it was decided to simplify the burn schedule by having only a single LOI burn and then, on LOI day, place the CSM/LM into the pre-descent orbit of about 60 by 7.5 n. miles. The LOI burn would result in an orbit of about 60 by 170 n. miles – the same as before except LOI-1 terminology was dropped. Then the DOI burn would achieve the combined objectives of the old LOI-2 and DOI - that is, on LOI day it would bring the spacecraft into the 58.5 by 7.5 n. mile pre-descent orbit, which would then precess to the desired 58.6 by 7.8 n. mile orbit at the time of PDI - about 1 day later.

Lunar Descent and Landing

Approximately 20 hours after the LOI-2 burn (for Apollo 11 and Apollo 12) the crew has rested and LM systems activation and checkout has been completed. The CSM then undocks from the LM and does station keeping to visually inspect the LM landing gear and structure, in preparation for the Lunar landing maneuvers.

First is the Descent Orbit Insertion (DOI) burn, which takes place behind the moon, a LM PGNCS controlled retrograde burn of about 30 seconds long (71 fps) using the LM Descent Propulsion System (DPS). DOI is a Hohmann transfer maneuver resulting about one hour later (one-half revolution later) in the LM descending to a 50,000 foot perilune that is 14 degrees prior to the targeted landing site.

Powered Descent Initiation (PDI) is the start of the DPS burn which begins near the 50,000 foot perilune and ends about 12 minutes later with LM lunar contact and DPS shutdown.

Apollo 12's Precision Landing

An important Apollo 12 objective was "development of techniques for precision landing capabilities" - to be demonstrated by designing the Apollo 12 trajectory for a LM landing within walking distance of the Surveyor III spacecraft which had been on the moon since April 1967 – 31 months earlier. This would also enable astronaut retrieval of Surveyor parts for later study of the effects of long-term exposure to the space environment.

The DPS DOI burn behind the moon placed the LM into a 9 x 69 nm orbit, which was followed about 1 hour later by a second DPS burn of 12-minutes within sight of the earth, the final moon landing burn.

Apollo 12 mission highlights posted by NASA states: "a newly-developed 'Lear' powered-flight data processor in Houston" used MSFN tracking data to calculate a landing site bias correction of 4,190 feet, which was voiced to the crew and entered into the LM computer about 2 minutes into the 12 minute powered descent burn. This was a complete success; NASA labeled Apollo 12 "the pinpoint mission" as the LM had landed about 535 feet from the Surveyor III - "closer to the target than expected."[14]

The contributions of TRW's Dr. William Lear to the Apollo program were many. His "Lear Filter" work not only was used to differentiate between PGNCS and AGS subtle failures during LM powered ascent from the moon, but also enabled the Apollo 12 LM to land within walking distance of the Surveyor III spacecraft. These filters basically processed ground station tracking data to more accurately determine the spacecraft trajectory.

---

[14] Apollo 12 Mission Highlights - dated Aug 3, 2017.
www.nasa.gov/mission_pages/Apollo/missions/apollo12.html.

MSC's Emil Schiesser, former MPAD Math-Physics Branch Chief, states in his 2006 oral history interview:

"Bill Lear from TRW started to help us. I asked him to work on the development of Kalman filters for the various Apollo navigation tasks. He was a really smart guy and easy-going; smoked a pipe, professor type, Dr. William Murphy Lear.

From then on and throughout all of the other programs he was the one we relied on for all our Kalman filter formulation and design. He could do more work in two months than a team of five people could do in six, and it would be better. This might be a bit of an exaggeration. But then he tended to work day and night. The Kalman filters developed by Bill are sometimes referred to as Lear Filters or sometimes the Lear filter, though there were more than one."[15]

---

[15] From Emil Schiesser's oral interview on 2 Nov 2006:
https://www.jsc.nasa.gov/history/oral_histories/participants_full.htm

Lunar Ascent & Rendezvous with the orbiting CSM

The ascent phase of the lunar landing mission begins when the LM lifts off the lunar surface, after GNC alignments, state vector updates, etc. are done as documented in the Lunar Surface Phase Mission Techniques document.

The nominal ascent trajectory consists of a vertical rise for about 10 seconds during which the LM is yawed to the proper azimuth and then pitched along a flight path targeted to insert the vehicle in a 9 by 45-nautical mile lunar orbit in-plane with the CSM's orbit.

During this time, the ground (MCC-H) is responsible for detecting and isolating insidious GNC systems malfunctions which would result in either a pericynthion altitude less than 30, 000 feet or wedge angle greater than 1.0 degrees (the angle between the CSM and LM orbital planes).

On reaching either or both of these limits, MCC-H would recommend the required action to the LM crew – switch from PGNCS to AGS or keep with PGNCS as AGS has failed, or another recommendation. The ground monitoring procedures were designed to protect against system errors which would violate these two constraints, and at the same time minimize the possibility of switching from a good PGNCS to AGS. In most instances, the PGNCS malfunctions required to reach a switchover limit greatly exceed the 3-sigma expected GNC systems performance.

Following ascent by the LM into a safe lunar orbit (nominally 9 by 45 nautical miles), the rendezvous maneuver sequence begins - CSI, CDH, TPI, then TPF.

Approximately 51 minutes after LM orbit insertion, the Concentric Sequence Initiation (CSI) maneuver is performed using the LM RCS. This nominally 50-feet per second maneuver results in a near-circular orbit of approximately 45 nautical miles to establish the proper phasing and differential altitude conditions for the CDH maneuver point. A plane change maneuver which aligns the LM and CSM orbital planes is executed (if required) 29 minutes after CSI.

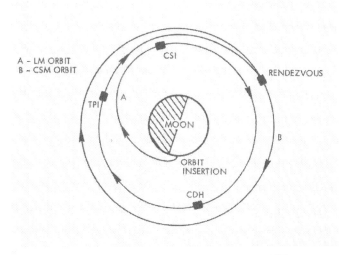

A – LM ORBIT
B – CSM ORBIT

CSI

RENDEZVOUS

TPI

A

MOON

B

ORBIT
INSERTION

CDH

COELLIPTIC RENDEZVOUS SEQUENCE

The Concentric Delta Height (CDH) maneuver of about 6 feet per second is executed one-half of the LM orbital period (about 58 minutes) after CSI and is designed to place the LM and CSM into co-elliptic orbits with a constant delta height of 15 nautical miles between the two orbits.

When the elevation angle from the LM to the CSM reaches 26.6 degrees (approximately 38 minutes after CDH), the LM Terminal Phase Initiation (TPI) maneuver is executed. This maneuver (nominally 26 feet per second) is designed to place the LM on a trajectory that will intercept the CSM orbit after 130 degrees of CSM travel. Two midcourse correction maneuvers are scheduled between TPI and braking: the first, 15 minutes after TPI, and the second, 30 minutes after TPI.

Terminal Phase Final (TPF) braking is initiated about 40 minutes after TPI, which circularizes the LM orbit at 60 nautical miles altitude. Docking is accomplished after completion of the braking maneuvers in which relative range and range rate are reduced to zero.

Following the rendezvous and docking, the LM crew configures the LM for jettison and then transfers to the CSM. The CSM executes a 1-foot per second retrograde maneuver after LM jettison to provide separation distance for the Transearth Injection burn.

George M. Low's Perspective

The March 1970 issue of Astronautics & Aeronautics contains a series of articles entitled "What made Apollo a Success" as shown here. Each article was written by a key MSC leader in the Apollo Program as shown here.[16]

Following are excerpts from George M. Low's INTRODUCTION that provide his perspective on the mission techniques/data priority activity. Mr. Low was Manager of the Apollo Spacecraft Program Office (ASPO) during Tindall's tenure.

---

[16] *The complete document can be downloaded at:*
https://ntrs.nasa.gov/archive/nasa/casi.ntrs.nasa.gov/19720005243
.pdf

1. INTRODUCTION

By George M. Low

Manned Spacecraft Center

On July 20, 1969, man first set foot on another planet. This "giant leap for mankind" represented one of the greatest engineering achievements of all time. This article and the others in this document describe and discuss some of the varied tasks behind this achievement."

We will limit ourselves to those tasks that were the direct responsibility of the NASA Manned Spacecraft Center: spacecraft development, mission design and mission planning, flight crew operations, and flight operations. We will describe spacecraft design principles, the all-important spacecraft test activities, and the discipline that evolved in the control of spacecraft changes and the closeout of spacecraft anomalies; and we will discuss how we determined the best series of flights to lead to a lunar landing at the earliest possible time, how these flights were planned in detail, the techniques used in establishing flight procedures and carrying out flight operations, and, finally, crew training and simulation activities -- the activities that led to a perfect flight execution by the astronauts.

In short, we will describe three of the basic ingredients of the success of Apollo: spacecraft hardware that is most reliable, flight missions that are extremely well planned and executed, and flight crews that are superbly trained and skilled. (We will not discuss two equally important aspects of Apollo -- the launch vehicles and launch operations. These elements, the responsibility of the NASA Marshall Space Flight Center and the NASA Kennedy Space Center, go beyond the scope of this series of articles.)

Mission Planning and Execution

Once basic missions had been defined, each flight had to be planned in detail. The mission planner tries to fit into each flight the maximum number of tests of the hardware and the widest variety of operations. For example, he will develop a rendezvous profile for a single earth-orbital flight that involves all of the normal and abnormal rendezvous conditions which might be encountered around the moon rendezvous from above, rendezvous from below, rendezvous with the lunar module active, rendezvous with the command and service module active, and rendezvous with varying lighting conditions. At the same time, the mission planner will try to exercise all of the propulsion systems and all of the navigation systems on both spacecraft.

After mission plans come the mission techniques (by another name, data priority). Given two or three data sources (for trajectory control), which of the sources should be believed and which discarded? Limits for each system had to be determined, *and logic flows for every conceivable situation had to be developed.*

Finally, the flight controllers take over. They had participated, of course, in the mission-planning and mission-technique activities; but now they had to work out each step of the flight and anticipate every emergency situation that might arise. What is the proper action when one fuel cell fails? What if two fail? The answers to thousands of questions like these had to be derived in terms of the specific mission phase. A rendezvous radar failure before command and service module-lunar module separation dictates that the two vehicles should not be separated. The same failure after separation allows the mission to be continued because the risk of rendezvous without radar has already been incurred and will not increase in subsequent mission sequences. Each of these events was documented as a mission rule long before the flight, and mission rules were placed under "configuration control," as was every other aspect of the Apollo system.

Flight controllers also worked out the best formats for their real-time displays. During the Apollo 11 descent to the surface of the moon, the flight controllers could watch, with a delay of only 6 to l0 seconds, the functioning of nearly every onboard system. They saw the rise in chamber pressure as the descent engine was throttled up to full thrust, and they could determine that the throttle-down occurred at the proper time. The flight controllers could also compare the descent trajectory from three data sources -- two onboard guidance systems and the ground tracking system. With this information, a flight controller on the ground could tell the crew, nearly 250 000 miles away, to ignore the alarms from the onboard computer during the most critical portion of the descent, because the system was guiding the spacecraft correctly.

Many of the techniques used during the flight were developed during countless hours of simulations. Simulation is a game of "what-if's." What if the computer fails? What if the engine does not ignite? What if... ? The game is played over and over again. The flight controllers do not know what situation they will face on the next simulation. By the time of flight, they will have done simulations so *often* and *they* will have worked together as a team so long, that they can cope with any situation that arises.

Because the Apollo equipment has worked so well and because there have been so few contingency situations, one could conclude that much of the planning, many of the mission techniques, and much of the training were done in vain. But this is an incorrect conclusion. As a minimum, the state of readiness that evolved from these efforts gave us the courage and the confidence to press on from one mission to the next. Also, there were situations -- the computer alarms during the descent of Apollo 11 and the lightning discharge during the launch of Apollo 12 -- that might have led to an abort if the team had been less well prepared and less ready to cope with the unexpected.

MSC Staff Comments

The following paragraphs are taken directly from the NASA/JSC Oral History Project.

Phil Shaffer – Flight Dynamics Officer (FIDO) - trajectory management for the early moon landing missions.

"The program office recognized that, and I don't remember the man's name that really surfaced it, but he did the program a favor and they invented the Apollo Data Priority, which was the precursor name for the mission techniques activity. They chartered Bill [Howard W.] Tindall [Jr.] out of the program office. He was a charging, assertive, integrating kind of guy, and they named him to lead the Apollo Data Priority."

Author's note. It was Marvin Fox who "did the program a favor, and invented Apollo Data Priority". From the beginning of the Tindall effort, Phil was a key individual representing Flight Control Division.

Shortly after the Apollo 11 mission Phil replaced Tindall as Chief of Apollo Data Priority Coordination - for the remaining Apollo missions.

Ronald Berry – MPAD Section Head & Branch Chief during Apollo.

"The purpose of the mission techniques was to say, "Look. You've got this trajectory that you want to fly to and you've got this hardware and software that you're going to use to attempt to fly to that trajectory, but how do you really do it? What targeting do you actually use? Who's primary for the targeting? Is it on board or is it ground? In whichever case that is, is the other one backup? How do they backup and monitor that? What are the actual parameters you use for monitoring? How big an envelope can you withstand before you have to change over from primary to backup?" This covered both guidance, navigation, incorporated the flight plan, what the crew's doing, what the ground crew's doing. It really was a systems integration of the entire program at that time. I'm not sure many people realize that that was a very critical, important activity."

"Bill and his mission techniques work and his famous Tindallgrams. I don't think the space program could have really pulled off what they did without his efforts and his activities."

Charles Deiterich – Retrofire Officer – responsible for abort/early lunar return situations.

"And Bill [Howard W.] Tindall had a group called the Data Priority Group, or Mission Techniques I think they later called it, but it was mostly the trajectory guys and the crew would get together and figure what we're going to do and how we're going to do it, what kind of stuff we're going to pass each other, how are you going to align the platform, where are you going to point it, what kind of errors can you take and still not have a problem, how much does a platform drift over time, how often do you have to do a platform alignment with the stars, do you turn the computer off while you're flying or don't you, do you leave it on? Do you turn the platform off? All those questions had to be answered."

"And, like I say, Tindall really did a very good job of pulling all that integration together. We probably wouldn't have made Apollo 8 happen had he not done those kind of things"

Glynn Lunney – Flight Director during the Gemini and Apollo programs

"Bill's team had developed a set of techniques for measuring what was happening to the orbits as we ran around the Moon well enough so that we could update the guidance as we went on down, in the descent phase, so that the lunar module knew where it was all the time and knew where it was relative to where we were trying to go—that is, the landing site." . . . "we found out that the Moon has these mass concentrations [mascons] close to the surface . . . they could perturb the orbit .....so, we had to develop techniques for sensing these perturbations"

Eugene Kranz – Flight Director during the Gemini and Apollo programs

"Now, there's one other honorary flight director; that was Howard W. "Bill" Tindall. Tindall was pretty much the architect for all of the techniques that we used to go down to the surface of the Moon."

Flight Director Gene Kranz said of Tindall "if there should have been a lunar plaque left on the Moon from somebody in Mission Control or Flight Control - it should have been for Bill Tindall. Tindall was the guy who put all the pieces together, and all we did is execute them."

Emil R. Schiesser – Chief, Orbit Determination in MPAD

That brings me to Tindall's "data priority" activity. Tindall was given the task of working out all of the flight techniques required for the execution of the Apollo missions around 1967.

Bill had to coordinate all of the factions of the program that were involved in the development and operation of the spacecraft, Mission Control Center, and other Apollo supporting ground facilities. Everything from hardware, software, the flight crew, the ground operations people, flight controllers, and all of the associated disciplines had an interest and input to be coordinated and resolved.

Tindall was very fair in conducting the meetings. He listened and respected all, belittling no one. The meetings would have anywhere from a few people up to perhaps seventy or more in them, depending on what was needed and how much exposure there had to be in real time as compared to after the fact. Of course, there couldn't always be a consensus. He would make the final decisions and though not all would agree, the decisions were respected. For example, if engineers could not agree on a number he might choose one if it was necessary. Upon the expression of concern by those present he would mention that all should go with that number and at such time as it was determined that it was not acceptable to come back with a better value and a good reason to change it and it would be changed. He knew that the activities and analysis in place would result in an assessment of the number and by choosing or defining its value, with a degree of wisdom, he kept the program moving forward.

If it weren't for Tindall we would never have landed on the Moon before the end of the decade, as requested by [President John F.] Kennedy.

Author's note: Before coming to MSC in 1962, Emil was part of the NASA Space Task Group at Langley Field, VA on Aerodynamic Theory and Numerical Analysis Techniques.

Marlowe Cassetti – Flight Operations Branch Chief during Apollo

His comments about "work"

So one time I brought my son in. He was either in junior high or high school, early high school or late junior high, and I thought, "Well, I'll show him what I do," so I snuck him into our building, and I gave him some graph paper, and we were getting the results off the computer, and he was plotting the results. We were trying to find out where the curve shifted, and I was explaining to him what this meant, what that meant to the Apollo Program, too. We got through—I mean, he was very serious about it. And he was into that kind of stuff. His favorite subject in school was physics and math.

So we got through with it, we were driving home, and he was real quiet, and I thought, "Well, maybe he's bored," so I said, "David, what did you think of all this? What did you think about this?"

He says, "I just can't believe it."

I said, "What can't you believe?"

He says, "That people really get paid for this. This is so much fun and they get paid for it." And that was exactly my reaction. I remember thinking back about when I saw this article about scientists doing this research and people really get paid for that.

Butler: That's really neat, especially for him to think that of what you're working on.

Cassetti: Yes. And now he's an engineer, a chip designer for Intel Corporation.

Bill Tindall's comments about "work"

From the NASA publication "Monographs In Aerospace History-Number 14 "Managing the Moon Program: Lessons Learned from Project Apollo" - Tindall states similar thoughts indicative of his outgoing and positive personality:

"But the thing that was so outstanding, you just hope that the young engineers and scientists that we are talking to here have a chance to be, to get into an organization, I don't know whether it has to be a project like Apollo, but an organization like we had that really delegated the jobs as tough or tougher than you could do and just said go on out there and figure out how you can do it because it was so doggone much fun."

"I mean, when Chris (Christopher Kraft) was talking, when he started out talking about these terrible 14 and 16 hour days and all during dinner that you were working, but I would just change the word from work to play because I never thought we were working at all. And that is the honest to God truth. It was just so much fun. In fact, I think it would be terrible if you had to go through life working. Really."

Mission Techniques beyond Apollo?

On January 5, 1972 President Richard M. Nixon announced that NASA would proceed with the development of a reusable low cost space shuttle system.

The Space Shuttle Orbiter with its crew of seven astronauts was launched like a rocket, orbited like a spacecraft, and landed like a plane. It carried satellites, space probes, and other cargo into orbit around Earth on both commercial and non-commercial missions.

The Shuttle program with its five orbiters (Columbia, Challenger, Discovery, Atlantis and Endeavor) completed a total of 134 successful missions over a 30 year period - from 1981 to 2011.

Two catastrophic Shuttle failures

After 24 successful shuttle flights, in 1986 Challenger was lost on its 10th flight, disintegrating 73 seconds after launch due to a booster O-ring failure. This resulted in a 2 ½ year suspension of shuttle flights.

In 2003, there was a second catastrophic failure when Columbia disintegrated during re-entry due to damage to its thermal protection system by a piece of foam insulation which broke off from the external tank during launch.

After this failure a designated rescue mission was ready if an on-orbit inspection of the thermal protection system prior to reentry revealed an irreparable damage.

Space Shuttle Mission Techniques?

About 7 years before the first Shuttle flight, Mr. Tindall wrote in 1974 the following memo providing rationale for why a Mission Techniques type effort should be done for the Space Shuttle program.

Whether or not this was done I do not know - but I thought it interesting.

NATIONAL AERONAUTICS AND SPACE ADMINISTRATION
LYNDON B. JOHNSON SPACE CENTER
HOUSTON, TEXAS 77058

$FILE$
$XXI-9$

REPLY TO
ATTN OF: FA-190

XXI-9 - Memo, JSC/NASA, March 6, 1974
SHUTTLE REDUNDANCY MGMT-ESPECI-
ALLY DURING CRITICAL MISSION
PHASES

MEMORANDUM

TO: Distribution

FROM: FA/Director of Data Systems and Analysis

SUBJECT: Shuttle Redundancy Management--Especially During
           Critical Mission Phases

This memorandum is to put on paper some thoughts with regard to
Shuttle redundancy management. I don't pretend to be an expert
in all Shuttle systems, but I think I know enough to state with
confidence that redundancy management has got to be one of, if
not the most challenging design problem in the Shuttle.

The specific problem I am speaking of, of course, is the manner
in which failures are detected and handled, particularly during
critical mission phases. The problem is very similar to the one
that we had on Apollo which resulted in the automatic EDS; that
is, the need for implementing an automatic system to perform a
crucial function because the crew's reaction time is not fast
enough. And like the EDS, I am sure every effort will be made
to disable the automatic redundancy management system as much of
the time as possible. The reason for this is that implementing
an automatic redundancy management system is really an attempt to
pre-program a man's judgment such that the system will do the
right thing in reconfiguring for all situations which unfortu-
nately includes those we cannot imagine. And that job, in my
opinion, is impossible, at least for the first several years of
Shuttle flights. In other words, it ain't perfect. And if it
ain't perfect, we're gonna want to turn it off during critical
mission phases whenever we can. If you will accept all of the
above, you must come to the conclusion that there will have to be
two basic modes for redundancy management during critical mission
phases--auto and manual. I am certain that there will be a significant proportion of the implementation which is common to both
of these modes. But I am equally certain there will be a substantial percentage which is not at all common in that the cues,
i.e., information which the crew needs to make their judgment,
will more often than not be different than the automatic system

would use. And, furthermore, the action that they perform may
very well be different than that which the automatic system is
programed to do.

So what. A gang of us--RI, IBM, MIT, JSC--more or less agree that
what I have written above fairly well describes the situation. At
least I think we all agree. And we also agreed that since the
activity associated with defining redundancy management requirements and its effect on subsystem design is a difficult and
schedule-critical business, we are aiming to do the following:
During the same week as the crew station review at Downey on
April 1, we will also conduct a day-long session on this subject.
(Somewhat loosely coupled to that, we will also spend a day on the
MIT Program Development Plan meeting; i.e., at least 3 days at
Downey.) The specific things we intend to do during this day-
long session are to review the work that Rockwell has been doing
and is having done to understand both the situation from a technical standpoint and also from the organization and management
standpoint. For example, we are specifically interested at what
level in the organization decisions of various criticality are
made as well as how the work of the far-flung organizations involved is integrated together and they are all kept up to date.
We will also review the work past, present, and future leading to
the decisions as to when the automatic redundancy management system must be enabled and when the crew can disable it and exercise
their superior judgment.

Finally, I guess I must confess that, at least in the manual mode,
the activity appears to smack heavily of our old friend "Mission
Techniques." This seems to be the case because what we are really
talking about is the comparison and management of various systems,
their characteristics, and the specific parameters and limits used
as cues to reconfigure as well as mission rules, crew procedures,
and all of the analyses that are carried out in their development.
What I am saying is I wouldn't be surprised if coming out of our
meeting we will all agree that something more or less equivalent
to the Mission Techniques meetings should be started. And I would
like to emphasize at this point that I am not just talking about
guidance, navigation, and control type functions, but other
critical systems as well such as electrical power.

So that this memorandum won't be misunderstood, I would like to
clearly state that I am by no means criticizing what has been
going on because I have no reason to do so. My point is that
there appears to be a need for the start of a new activity which

probably couldn't have begun much sooner than this in any case; but that now that the basic design of the various systems is jelling, it becomes appropriate to start figuring out how they should be used.

Howard W. Tindall, Jr.

Distribution:
CA/K. S. Kleinknecht
  E. F. Kranz
  J. W. Roach
  W. J. North
CB9/C. M. Duke
  J. W. Young
EA/M. A. Faget
EA2/R. A. Gardiner
EJ/W. C. Bradford
EJ3/K. J. Cox
LA/R. F. Thompson
MA/A. Cohen
  M. A. Silveira
ME/R. W. Kubicki
MIT, N. E. Sears
  S. L. Copps

FA/L. C. Dunseith
  R. G. Rose
  J. A. Frere ✓
  C. C. Critzos
FA2/R. E. Ernull
FD/D. E. Iloff
FE/C. H. Woodling
FM/J. P. Mayer
FR/R. P. Parten✓
FS/J. C. Stokes, Jr.

cc:
AA/C. C. Kraft, Jr.
AB/S. A. Sjoberg
AC/G. W. S. Abbey

FA/HWTindall,Jr:jsw:3181   3/5/74

IN CONCLUSION

I was fortunate to not only have been the TRW lead engineer for the pre-Tindall work, but also managed the TRW efforts under him for the next 2 years – until the first lunar landing.

Our TRW task group was in effect Bill's "staff" by creating the draft documents and the logic flow diagrams in the documents using inputs from the working group participants. We conducted reviews/revisions within the various MSC working groups, some without Bill, and after a final review and concurrence by the working group I would meet with him for his signature on the cover letter. Then we published and delivered to MSC the many copies needed for his extensive distribution list.

One cannot say enough about Bill Tindall in his role as Chief of Apollo Data Priority Coordination. Although TRW identified the need for this effort and initiated their development, it was Bill who very effectively managed their subsequent development for the next 2-plus years - up until the first lunar landing. His responsibilities were wide ranging; he chaired many meetings between astronauts, mission controllers, design engineers, contractors, and other relevant parties, adjudicated disagreements and oversaw the details of the mission techniques effort. Many agree he was one of the most important contributors to the success of the Apollo program.

Along the way Bill also wrote his "Tindallgrams" documenting his meetings and explaining the rationale for the many decisions made along the way. These are very interesting to read – his writing style was characteristic of his inclusive personality and ability to clearly communicate complex concepts.

Finally Marvin Fox, my TRW supervisor during those times. Early in 1966 Marvin recognized the need for a documented and controlled baseline plan for trajectory control that would include the real-time decision logic for GNC data verification and maneuver determination. He called this Data Priority, and in early 1966 enthusiastically promoted this need and concept to key NASA individuals. The initial effort for Dr. Shea created the first Apollo Mission Techniques documents covering the two most critical phases of the mission - lunar descent onto the moon and ascent/rendezvous with the CSM "mother ship" orbiting the moon.

Marvin today lives in Houston not far from NASA/JSC.

APPENDIX

BRIEF DOCUMENT SUMMARIES

This appendix contains short summaries of the eight mission techniques documents shaded in the table below to illustrate their scope and level of detail - which may not be readily apparent to the reader.

These documents were those in effect at the time of Apollo 11 flight.

DOCUMENT	MSC REPORT	UHCL #
1. Launch Phase Aborts	SP-PA-9T-048	37285
2. Earth Parking Orbit and Translunar Injection	SP-PA-9T-144	480695
3. Translunar MCC's, Lunar Orbit Insertion (and 6/10/69 change pages)	SP-PA-9T-41	37691 480689
4. Lunar Orbit Activities	SP-PA-9T-135	480694
5. Lunar Descent	SP-PA-8N-021A	480691
6. Descent Abort and Subsequent Rendezvous	SP-PA-9T-137	480693
7. Lunar Surface Phase	SP-PA-9T-050	208074
8. Lunar Powered Ascent	SP-PA-9T-128	480692
9. Transearth Injection, MCC's and Entry	13. SP-PA-8T-028	35516
10. Tracking Data Select Controllers Procedures	SP-PA-9T-049	37398
11. Contingency Procedures	SP-PA-9T-043	37228
12. Manual Ascent	SP-PA-9T-055	480696

1. Launch Phase Aborts

The mission time covered by this document is only 12 minutes, the time from liftoff of the giant Saturn V launch vehicle with its Apollo Spacecraft payload and the attainment of a safe Earth Parking Orbit.

This document describes the planned sequences for terminating the mission in the event of critical launch vehicle or spacecraft malfunctions which preclude continuing the mission. The document includes MCC-H flight controller and flight crew procedures and decision logic needed for each non-nominal situation that could be encountered during this time period.

The Saturn V launch vehicle (LV) is a four stage booster designed to place the Apollo spacecraft (SC) onto a translunar trajectory. There are three powered stages: the S-IC, powered by five F-1 engines; the S-II, powered by five J-2 engines; and the S-IVB, powered by a single J-2 engine. The unpowered Instrument Unit (IU) is the fourth stage. It rides between the S-IVB and the spacecraft and contains Saturn V's GN&C system.

A normal launch consists of a boost into Earth Parking Orbit (EPO) by burning practically all the S-IC and S-II stage propellants and partially burning the S-IVB propellant. A second S-IVB burn, the translunar injection (TLI) burn, is done after 2-4 revolutions in the EPO.

Figure 2-1 Mission F Launch Timeline and Contingency Capability (taken from this document) summarizes the contingency capabilities available during a typical Mission G launch. Abort and contingency maneuvers described in the document are:

a) Automatic aborts during S-1C flight using the Launch Escape Tower.

b) Early staging of the S-IVB stage from the S-II stage in the event of S-II stage malfunctions which preclude its continued powered flight.

c) Launch Vehicle malfunctions wherein one or more posigrade SPS burns would result in achieving an orbit having a perigee altitude of at least 75 nautical miles – a "safe" orbit.

d) Apogee Kick - a posigrade SPS burn at apogee to achieve a safe orbit.

e) Manual take-over by the crew during S-II or S-IVB flight, but only if onboard cues verify that the IU platform has failed.

f) The Mission Techniques for Contingency Procedures states: "The crew has the capability to manually steer the TLI burn if the IU platform has failed. but adequate data are not yet available to determine if it is wise to do so, knowing the IU platform has failed prior to TLI. Until this question is resolved. TLI is no go if the IU platform has failed."

Thanks to Von Braun and his outstanding engineering team at Marshall Space Flight Center (MSFC), none of these contingency capabilities were required as all eleven Saturn V launch vehicles used for the manned Apollo missions performed successfully.

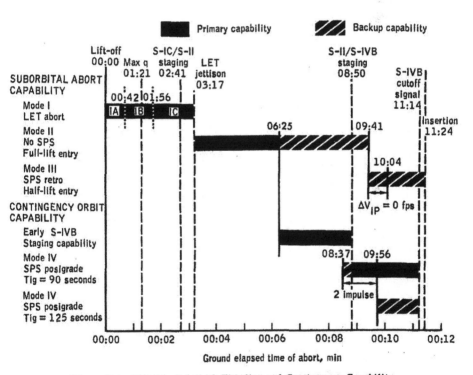

Figure 2-1. Mission F Launch Timeline and Contingency Capability

2. <u>Earth parking orbit and Translunar Injection</u>

The Saturn V launch vehicle placed the Apollo CSM/LM spacecraft assembly into a nominal 115 n. mile circular earth orbit. Then nominally after about one-and-a-half revolutions the Saturn V S-IVB stage is ignited for a second burn, the translunar injection (TLI) burn of 5-6 minutes to place the Apollo spacecraft into a translunar orbit.

This document covers two important ground rules affecting the TLI GO/NO GO decision. These are:

1) A properly operating CSM GNCS is mandatory for TLI. Section 3 in this document describes the various monitoring checks and comparison limits used to determine if the PGNCS is a go for TLI.

2) The TLI maneuver will not be attempted using the Saturn V IU if it is not working properly, as described in Section 2 of this document.

For Saturn V IU accelerometer failures prior to the TLI burn, the Huntsville Operations Support Center will determine the probability that the IU can still guide the booster into a 65,000-nautical mile apogee orbit. If that probability is 95 percent or more, TLI is go.

The procedure for manual take-over (MTO) of booster steering by the crew is also shown if the IU has failed and the PGNCS is OK. This procedure includes an ORDEAL altitude setting.

EPO procedures covered in Section 4 include a Stabilization and Control System (SCS) drift test, an optics go/no go status check, an optional IMU realignment, a MSFC S-IVB guidance and fuel reserve go/no go decision, a CMC state vector update, and finally TLI preparations and a burn attitude check. Section 5 covers TLI burn monitoring.

The final 10 pages contain the logic flow diagrams for the EPO activities and decisions discussed in Sections 4 and 5.

3. Translunar Midcourse Corrections & Lunar Orbit Insertion

The techniques presented herein reflect the following ground rules:

- MCC-H will always provide maneuver targeting for midcourse corrections and LOI-1 and LOI-2.

- All burns will use the GNCS external ΔV mode.

- MCC-3 is the focal point for translunar midcourse corrections. Nominally, MCC-3 will place the spacecraft on a trajectory which should not require additional MCC.

- If possible the SPS will be used for all midcourse corrections to conserve RCS fuel.

- A properly operating GNCS is mandatory for LOI.

Typical LOI go/no-go criteria, from this document, that MCC-H will evaluate during translunar coasting are:

- accelerometer bias. If sufficiently large to suspect that failure is evident (greater than. O. 164 foot per second squared) then LOI is no-go.

- gyro drift errors. If this error indicates that a GNCS failure is evident (greater than 1. 5 degrees per hour) then LOI is no-go. If the drift error is greater than 0.075 degree per hour, then gyro updates are performed.

- If other data comparison limits are exceeded or unfavorable trends exist, then a temporary LOI no-go is given until the orbital decision parameters have been investigated. It is not necessary for MCC-H to evaluate the orbital decision parameters (differences between GNCS and MSFN tracking) unless the strip chart limits are violated. If the orbital parameter differences are within limits, then LOI is go at this point.

- The state vector comparison display will be used to evaluate GNCS performance after each midcourse correction. If the trajectory is determined to have been significantly perturbed and if it is determined that a GNCS failure is the cause of the perturbation, then LOI is no-go.

4. **Lunar Orbit Activities**

In this document, the lunar orbit activities are divided into two distinct phases.

The first phase starts when the lunar parking orbit circularization maneuver (LOI-2) is completed and consists of the following sequence of events:

- LM partial activation – about 2 hours
- Crew rest period – about 9 hours
- LM activation and systems checkout – start of "DOI day"
- Undocking of the CSM from the LM
- Station keeping by the CSM to inspect the LM landing gear and structure
- CSM separation maneuver
- LM descent orbit insertion (DOI) maneuver

This phase ends just prior to the initiation of powered descent (PDI).

The second phase starts at the completion of powered ascent and consists of the following sequence of events:

- LM active rendezvous (CSI/CDH/TPI/TPF)

- LM/CSM docking

- LM jettison

This phase ends just prior to initiating preparations for the TEl burn.

First Phase – Preparation for Powered Descent Initiation

Following LOI-2 the LMP and CDR enter the LM and perform the initial LM activation. The LM GNCS are not powered at this time. During this period, the CMP performs a CSM IMU to the landing-site REFSMMAT uplinked earlier, and pseudo landing site observations to provide real-time training and familiarization with the terrain.

Following the 9-hour sleep period, the LMP and the CDR reenter the LM and complete the LM activation and systems checkout. This includes state vector and REFSMMAT updates from MCC-H, and IMU alignments and drift checks, for all GNC systems. Landing-site observations are again performed by the CMP one revolution prior to separation, and the data are processed by the MCC-H to obtain a better relative position of the landing site with respect to the orbit.

After the CSM undocks from and inspects the LM, the CSM performs a 2.5 fps radial inward RCS separation burn that results in the CSM leading the LM by 13, 100 feet at DOI one-half revolution later.

The DOI maneuver is performed at about 194 degrees prior to the targeted landing site and is a PGNCS controlled retrograde burn of about 71 feet per second using the DPS. This Hohmann descent transfer maneuver serves to reduce perilune to 50,000 feet about 14 degrees prior to the targeted landing site, the position at which powered descent is initiated.

Second Phase – Rendezvous, Docking and LM Jettison

The LM active rendezvous begins just after LM ascent orbit insertion into a nominal 9 by 45 nautical miles orbit, timed so the CSM is leading the LM by approximately 16 degrees central angle.

The CSI maneuver of about the 50 fps is performed about 51 minutes after APS orbit insertion using the LM RCS. This maneuver results in a near-circular orbit of about 45 nautical miles and establishes the proper phasing and differential altitude conditions for the CDH maneuver point.

If required a plane change maneuver that aligns the LM and CSM orbital planes is executed about 29 minutes after CSI.

The CDH maneuver of about 6 feet per second is executed one-half of the LM orbital period later (approximately 58 minutes after CSI) and is designed to place the LM and CSM into co-elliptic orbits with a constant delta height of 15 nautical miles between the two orbits.

When the elevation angle from the LM to the CSM reaches 26.6 degrees (approximately 38 minutes after CDH), the LM terminal phase initiation (TPI) maneuver is executed. This maneuver (nominally 26 feet per second) is designed to place the LM on a trajectory that will intercept the CSM orbit after 130 degrees of CSM travel.

Two midcourse correction maneuvers are scheduled between TPI and braking: the first, 15 minutes after TPI, and the second, 30 minutes after TPI.

Terminal Phase Final (TPF) braking is initiated about 40 minutes after TPI, which circularizes the LM orbit at 60 nautical miles altitude. Docking is accomplished after completion of the braking maneuvers in which relative range and range rate are reduced to zero.

Following the rendezvous and docking, the LM crew configures the LM for jettison and then transfers to the CSM. The CSM executes a 1-foot per second retrograde maneuver after LM jettison to provide separation distance for the TEl burn. The remainder of the lunar orbit activities are devoted to TEl preparations.

9. Transearth Injection, Midcourse Corrections and Entry

Similar to the time period from TLI to LOI, the time from Transearth Injection (TEI) to entry is about three days. Midcourse correction and entry targeting had basic ground rules similar to those for the translunar and LOI techniques:

- The ground is prime for all maneuver targeting.

- All burns will use the GNCS External ΔV mode.

- The midcourse corrections will normally be targeted for corridor control only.

- If a total communications failure occurs, the onboard navigation system will be used to effect a safe entry.

- the entry range shall be limited to from 1200 to 1800 nautical miles. That is, the nominal entry trajectory profile will not include a skipout phase.

10. Tracking Data Select Controllers Procedures

The tracking Data Select Controller is responsible for the determination of the best state vector to be used for all trajectory control decisions and information. In addition, he advises other mission controllers of the state vector quality, as well as the quality of the data sources when possible or when so requested.

This document contains much detailed information and procedures concerning the Tracking Data Selection Controller's responsibilities in the RTCC operations. *Data Select* is this position's identifier.

Described also are this person's communication interfaces with:

RTCC Supervisor – the individual responsible for overall RTCC operations.

Track – the individual responsible for the receipt of tracking data from GSFC and input of the data to the real-time system.

Lemon One – the person located at the Cape responsible for the selection of the impact prediction data source.

FDO – the flight controller responsible for all dynamic operations and decisions.

Goddard Analyst – responsible for preliminary analysis of the tracking data received by Goddard.

Select Support – an advisor to *Data Select* who is also responsible for cislunar and lunar tracking schedule planning.

The first 24 pages cover these topics as well as tracking site calibration requirements, and tracking site selection criteria as needed to establish the best station-to-station geometry for the purpose of orbit determination.

Finally, detailed procedures for each phase of the lunar mission with supporting logic flow diagrams are contained in the final 28 pages of the document.

11. Contingency Procedures

This document defines the GNC monitoring techniques and resultant contingency procedures for all types of failures that involve real-time decisions by the crew and ground (MCC-H) - during six mission phases and for the Navigation function when communications between the MCC-H and the spacecraft are lost.

Launch Phase. This short section summarizes and references the Launch Phase Aborts Mission Techniques Document.

EPO. Earth Parking Orbit contingency procedures consist of monitoring critical SC systems and those LV systems required for TLI in order to determine whether or not to proceed with the TLI burn. In the event TLI is no-go, the following alternatives are available:

- Execution of potential alternate missions.

- Premature earth reentry at some time during EPO.

This short section references the Earth Parking Orbit and Translunar Injection Mission Techniques document; however, it also states:

"The crew has the capability to manually steer the TLI burn if the lU platform has failed. but adequate data are not yet available to determine if it is wise to do so, knowing the lU platform has failed prior to TLI. Until this question is resolved. TLI is no go if the lU platform has failed."

TLI Through Transposition, Docking and LM Extraction

(TD&E). The TLI burn is initiated during the second or third EPO revolution and nominally lasts for about 5 minutes.

The contingency procedures described in detail consist of:

- (1) TLI burn and postburn monitoring up to transposition and docking,

- (2) procedures for the 90-minute abort mode, TLI 10-minute abort mode, the 4-hour (no-voice) abort mode,

- (3) procedures for manual takeover for IU platform failure during TLI.

TLC (Translunar Coast) Detailed TLC abort procedures are presented which assume that an impending spacecraft system failure has been identified and that a TLC abort maneuver must be executed. All abort maneuvers are performed in either an external delta V mode or an SCS auto mode.

With communications, MCC-H will provide the necessary solution and specify the configuration and propulsion system to be used, depending on a real-time assessment of the contingency situation.

Possible return options are described such as a lunar flyby for a non-time-critical situation or minimum time returns targeted to any Contingency Landing Area (CLA). The entry velocity constraint of 37,500 feet per second defines the minimum return time.

LOI The document divides Lunar Orbit Insertion phase contingencies into the following four categories:

a) SPS related problems which can be identified by the nominal monitoring techniques (e. g., helium pressure),

b) Non-SPS problems which would not preclude completion of LOI but would either require an alternate mission or possibly an early TEl maneuver.

c) Inadvertant SPS shutdown - this includes all contingencies which can be identified with an SPS failure, however remote.

d) Guidance and control problems,

The document presents the SPS burn monitoring flow chart which defines the crew, real-time decision logic for use in monitoring the SPS, G&N, and critical spacecraft systems. In addition, it also defines the manual SPS shutdown logic when a contingency is identified by the monitoring procedure.

TEI The Transearth Insertion maneuver is a critical SPS burn in that it is mandatory for earth return. Therefore, the burn will be completed without shutdown if at all possible.

Like LOI, TEl occurs behind the moon and so the monitoring procedures and techniques are essentially the same for both maneuvers. However, TEl contingency procedures are presented for the unlikely occurrence of either (1) an inadvertent SPS shutdown or (2) an extremely critical situation which crew discretion dictates a manual SPS shutdown.

NAVIGATION. The basic ground rule is for MCC-H to provide the navigation function as long as communications are maintained with the spacecraft. If communications are lost, the navigation function must be performed onboard in accordance with the appropriate navigation schedule.

Since the LM provides a perfect communications backup during translunar coast, a navigation sighting schedule is not required. In the remote case that communications are lost, Navigation Sighting Intervals would be scheduled every 2 ½ hours and before and after each MCC in order to ensure a safe entry.

The section provides much detail here, including logic flow charts, typical block uplinks, typical sighting schedules and Sextant usage and constraints.

12. **Manual Ascent**

Manual ascent from the lunar surface will be necessary in the event of a double failure disabling automatic guidance steering in both PGNCS and AGS. While the probability of any specific pair of failures occurring is extremely small, there are many combinations of double failures which would disable both PGNCS and AGS automatic steering. Table 1 in the document shows these.

Also stated:

- For a safe manual insertion, it is necessary for the crew to have both an accurate attitude reference and a usable attitude control mode.

- Simulations have shown that the LM crew should be able to guide the LM into a safe orbit quite satisfactorily using the horizon viewed through the overhead window as an attitude reference. The resultant orbit can be far from nominal, which could present rendezvous problems.

- Both the AGS/CES (Control Electronics Section) and the PGNCS have a substantial capability, even if the accelerometers are broken. However, special procedures are required to use this capability.

- The failure of an attitude gyro in either guidance system virtually eliminates the use of that guidance system for vehicle control. The failure of a CES rate gyro causes loss of automatic capability in the disabled axis only.

Contingency procedures are presented for the following system failures:

- Failed AGS and failed PGNCS accelerometer

- Failed AGS and failed LGC

- Failed PGNCS and failed AGS X-axis accelerometer

- Failed PGNCS and failed AGS Z-axis accelerometer

- Failed PGNCS and failed AGS Y-axis accelerometer

- Failed PGNCS and failed CES rate gyro

In most manual insertion instances, a CSM active rendezvous will be used.

The LM's insertion orbit may have an apolune as great as 250 n. miles and a too-low perilune that requires the LM to perform an RCS burn at apolune. MCC-H will compute and voice up maneuvers for the LM as needed to ensure a minimum LM orbit of 16 by 30 n. miles prior to the CSM rendezvous maneuver sequence.

During the CSM active rendezvous, the CSM will perform the following maneuvers: phasing (NC1), CSI, CDH, and TPI. The phasing maneuver performed about one revolution after insertion establishes the proper phasing for the CSI maneuver point. The CSI maneuver, performed either 1/2 or 1-1/2 revolutions after NC1, establishes the proper phasing and differential altitude conditions for the CDH maneuver point. The CDH maneuver, performed 1/2 period after CSI, is designed to place the CSM into an orbit which is coelleptic with the LM's orbit with a constant delta height between the two orbits. The TPI maneuver is designed to place the CSM on a trajectory that will intercept the LM orbit after 130 degrees. Line-of-sight braking may be performed by either the LM and/or the CSM.

BIBLIOGRAPHY

Apollo 17 Flight Plan, Change A, NASA/MSC publication,1972

Apollo 17 Guidance and Navigation Summary General Motors Corp., Delco Electronics,1972

Challenge to Apollo: the Soviet Union and the Space Race, 1945-1974. Siddiqi, Asif A., NASA SP-2000-4408, 2000

Destination Moon: A History of the Lunar Orbiter Program - Chapter II. Langley enters the picture. NASA publication, https://history.nasa.gov/TM-3487/contents.htm, 1976

How Apollo Flew to the Moon, Woods, David W., Springer-Praxis Publishing, 2011

LM AGS Programmed Equations Document, Flight Program 6, TRW Systems Group, 1969

Managing the Moon Program: Lessons Learned from Project Apollo, Monographs in Aerospace History-Number 14, NASA publication July, 1999

Mueller, George E. biography. The Encyclopedia Astronautica, http://www.astronautix.com/m/mueller.html.

Phillips, Samuel C. biography The Encyclopedia Astronautica, http://www.astronautix.com/p/phillipssamuel.html.

Richard Boudreau Apollo Mission Techniques Documents. TRW Houston Operations publications, 1966

Techniques of Controlling the Trajectory. Tindall, Howard W. Jr., Astronautics & Aeronautics, March 1970

The Secret of Apollo - Systems Management in American and European Space Programs. Johnson, Stephen B., The Johns Hopkins University Press, 2002

TRW 1901 – 2001, A Tradition of Innovation TRW Inc., Cleveland, Ohio, 2001

TRW as MSC Support Contractor. Oral History Interview of Arnold Rosenbloom. JSC Staff Historian, Merrifield, Robert B., July, 1968

Two Sides of the Moon, Scott, David and Leonov, Alexei. St. Martin's Press, 2004

Von Braun, Dreamer of Space, Engineer of War, Neufeld, Michael J., Random House, 2007

NOTES

NOTES

NOTES

Made in the USA
San Bernardino, CA
07 June 2019